AF879104

Springer Tracts in Advanced Robotics 99

Editors

Prof. Bruno Siciliano
Dipartimento di Ingegneria Elettrica
e Tecnologie dell'Informazione
Università degli Studi di Napoli
Federico II
Via Claudio 21, 80125 Napoli
Italy
E-mail: siciliano@unina.it

Prof. Oussama Khatib
Artificial Intelligence Laboratory
Department of Computer Science
Stanford University
Stanford, CA 94305-9010
USA
E-mail: khatib@cs.stanford.edu

For further volumes:
http://www.springer.com/series/5208

Editorial Advisory Board

Oliver Brock, TU Berlin, Germany
Herman Bruyninckx, KU Leuven, Belgium
Raja Chatila, ISIR - UPMC & CNRS, France
Henrik Christensen, Georgia Tech, USA
Peter Corke, Queensland Univ. Technology, Australia
Paolo Dario, Scuola S. Anna Pisa, Italy
Rüdiger Dillmann, Univ. Karlsruhe, Germany
Ken Goldberg, UC Berkeley, USA
John Hollerbach, Univ. Utah, USA
Makoto Kaneko, Osaka Univ., Japan
Lydia Kavraki, Rice Univ., USA
Vijay Kumar, Univ. Pennsylvania, USA
Sukhan Lee, Sungkyunkwan Univ., Korea
Frank Park, Seoul National Univ., Korea
Tim Salcudean, Univ. British Columbia, Canada
Roland Siegwart, ETH Zurich, Switzerland
Gaurav Sukhatme, Univ. Southern California, USA
Sebastian Thrun, Stanford Univ., USA
Yangsheng Xu, Chinese Univ. Hong Kong, PRC
Shin'ichi Yuta, Tsukuba Univ., Japan

STAR (Springer Tracts in Advanced Robotics) has been promoted under the auspices of EURON (European Robotics Research Network)

Anh-Van Ho · Shinichi Hirai

Mechanics of Localized Slippage in Tactile Sensing

- and Application to Soft Sensing Systems

Springer

Anh-Van Ho
Department of Robotics
Ritsumeikan University
Shiga
Japan

Shinichi Hirai
Department of Robotics
Ritsumeikan University
Shiga
Japan

ISSN 1610-7438 ISSN 1610-742X (electronic)
ISBN 978-3-319-04122-3 ISBN 978-3-319-04123-0 (eBook)
DOI 10.1007/978-3-319-04123-0
Springer Cham Heidelberg New York Dordrecht London

Library of Congress Control Number: 2013956529

© Springer International Publishing Switzerland 2014
This work is subject to copyright. All rights are reserved by the Publisher, whether the whole or part of the material is concerned, specifically the rights of translation, reprinting, reuse of illustrations, recitation, broadcasting, reproduction on microfilms or in any other physical way, and transmission or information storage and retrieval, electronic adaptation, computer software, or by similar or dissimilar methodology now known or hereafter developed. Exempted from this legal reservation are brief excerpts in connection with reviews or scholarly analysis or material supplied specifically for the purpose of being entered and executed on a computer system, for exclusive use by the purchaser of the work. Duplication of this publication or parts thereof is permitted only under the provisions of the Copyright Law of the Publisher's location, in its current version, and permission for use must always be obtained from Springer. Permissions for use may be obtained through RightsLink at the Copyright Clearance Center. Violations are liable to prosecution under the respective Copyright Law.
The use of general descriptive names, registered names, trademarks, service marks, etc. in this publication does not imply, even in the absence of a specific statement, that such names are exempt from the relevant protective laws and regulations and therefore free for general use.
While the advice and information in this book are believed to be true and accurate at the date of publication, neither the authors nor the editors nor the publisher can accept any legal responsibility for any errors or omissions that may be made. The publisher makes no warranty, express or implied, with respect to the material contained herein.

Printed on acid-free paper

Springer is part of Springer Science+Business Media (www.springer.com)

To Our Beloved Family

Foreword

Robotics is undergoing a major transformation in scope and dimension. From a largely dominant industrial focus, robotics is rapidly expanding into human environments and vigorously engaged in its new challenges. Interacting with, assisting, serving, and exploring with humans, the emerging robots will increasingly touch people and their lives.

Beyond its impact on physical robots, the body of knowledge robotics has produced is revealing a much wider range of applications reaching across diverse research areas and scientific disciplines, such as: biomechanics, haptics, neurosciences, virtual simulation, animation, surgery, and sensor networks among others. In return, the challenges of the new emerging areas are proving an abundant source of stimulation and insights for the field of robotics. It is indeed at the intersection of disciplines that the most striking advances happen.

The *Springer Tracts in Advanced Robotics (STAR)* is devoted to bringing to the research community the latest advances in the robotics field on the basis of their significance and quality. Through a wide and timely dissemination of critical research developments in robotics, our objective with this series is to promote more exchanges and collaborations among the researchers in the community and contribute to further advancements in this rapidly growing field.

The monograph by Anh-Van Ho and Shinichi Hirai is based on the first author's doctoral thesis on the slippage of soft fingertips in tactile sensing. The contents of the volume are organized in two parts, each featuring three chapters, dealing with mechanical modelling and tactile sensing of localized slippage respectively. The original beam bundle model is introduced which allows a keen formulation of robotic object grasping and soft manipulation with haptic sensation. Interestingly enough, this concept can be extended from artificial to human fingertips and thus is useful in the analysis of human haptic sensation.

Rich in both numerical simulation and experimental validation, this volume is the second contribution to the series on the use of tactile sensing and as such it constitutes a very fine addition to STAR!

Naples, Italy
November 2013

Bruno Siciliano
STAR Editor

Preface

This book presents an approach for assessing slip perception of soft fingertips through dynamic simulations, and the application of this approach to stick-slip detection of robotic fingers. Much work in robotic dexterous manipulation has focused on methods of modeling slip motion, and considerable effort in of robotic anthropomorphic hands has centered on methods of stick-slip detection methods, along with embedded specific designs of sensory systems. Most of this work, however, involved quasi-static and analytic issues, with little introducing aspects of friction to the contact surface. To date, therefore, humans cannot use the proposed slip detection methods. To address remaining issues, we propose a hybrid model of a sliding soft fingertip, consisting of a virtual beam model and a Finite Element (FE) model. While the former can model normal and tangential deformations of the soft fingertip, the latter can model micro movements on the contact surface when sliding. We call this hybrid model the Beam Bundle Model (BBM). We also introduce dynamic friction into each contact node to fully describe sliding motion. In this research, we especially focus on the initial phase of movement, i.e. the pre-slide phase. We have formulated a Localized Displacement Phenomenon (LDP), which represents micro movements of contact points during pre-sliding of the object. This LDP is important in assessing stick/slip states and crucial in stable grasp/manipulation tasks. Utilizing this phenomenon, we have proposed several methods to detect incipient slippage of tactile fingertips using a fabricated micro force/torque sensor and robotic skins developed from special yarns and tactile-arrayed sensors. Finally, using LDP, we attempted to enhance tactile sensing so that it would be comparable in popularity to vision. Our approach can be generalized to slippage of other types of soft fingertips of known shapes and can be applied to haptic devices.

Localized slippage occurs during any relative sliding of soft contacts, ranging from human fingertips to robotic fingertips. Although this phenomenon is dominant for a very short time prior to gross slippage, localized slippage is a crucial factor for any to-be-developed sensing system to respond to slippage before it occurs. The content of this book addresses all aspects of localized slippage, including modeling and simulating it, as well as applying it to the construction of novel sensors with slip tactile perception.

One of the featured points of this proposal is to introduce a new way of looking at the sliding of soft fingertips as specific or soft objects. The proposed BBM suggested a method of modeling soft fingertips, in order to determine how and when slippage occurs during the sticking phase through the propagation of localized displacements on the contact area. As a result, the BBM helps readers intuitively imagine which phenomenon dominates each sliding contact characteristic, and validation experiments make the readers believe that it really exists. Robots and machines requiring sensing systems to look at and assess what is going on over the entire contact area, enabling them to respond to slippage in a timely way (see Part II, including Chapters 5 and 6). Thus, these sensing systems are considered undetachable. Localized displacement occurs during a very short period of time during the sticking phase, and it has been challenging to detect this phenomenon quantitatively. In Part I of this proposal, we employed a direct method, in which a high speed camera tracks movements at contact points, showing where the contact pad is sliding. Due to its size, this method, however, is not suitable for embedding in any tactile sensing system. Therefore, novel sensing systems, easy to fabricate and indirectly conforming to the LDP, are needed. Any indirect method must take into account localized displacements into its transduced output so that robots/machines can assess when slippage is about to occur.

In Chapter 5, we propose a tactile sensing system specialized for hemispherical soft fingertips. We developed a micro force/tactile sensor and embedded this sensor into the fingertip, in a position such that the fingertip conformed to the natural deformation of a human fingertip. This sensor obviously cannot directly sense dynamic changes of localized displacement, although simulations utilizing the BBM showed that signals from the sensor can predict movements of the central zone of the contact pad, movements that occurred immediately prior to overt slippage. Thus, thanks to BBM and LDP, the sensor can detect the incipient slippage properly without the necessity of observing LDP directly, indicating that analysis using BBM and LDP can result in a method of equipping the soft sensing system with slip perception.

While the sensing system in Chapter 5 was designed to detect the slippage of one zone on the contact pad, the developed textile sensor acts by transducing changes caused by localized displacements to its output. The specialized design of this sensor arose from analysis of localized displacement phenomena. This chapter has been included to show that our assessing system is simple but efficient in the detection of slippage. Chapters 5 and 6 help readers easily relate proposed BBM and LDP to practical applications, encouraging these readers to propose other novel tactile sensing systems. These chapters also suggest the important role of sliding action for the development of robots. For each specific tactile sensing system, a suitable method of decoding slippage in sensors' output (e.g. lag time t_d in Chapter 5, the discrete wavelet method in Chapter 6, and the slip indicator in Chapter 7) may enable ways to correctly detect slippage and perceive texture.

In deciding to publish a book, we considered how to draw a big picture, so that readers can imagine a completed slip, starting from modeling to practical application. Assessments of citations of our publications showed that researchers cited individual components of our work separately, i.e., researchers who worked on modeling only cited BBM-related papers, whereas those who worked on tactile systems cited papers describing the design of novel sensors. Neither set of researchers focused on the relationship between these two aspects of our work, with neither recognizing that the former was the foundation of the latter. As researchers working in a very practical field of robotics, we expect papers to mention both foundations and applications. If one is lacking, readers would hardly find them persuasive. Future readers of this book should take that into account, especially those working in tactile perception, a very wide aspect of research in robotics.

The contents of this book are based primarily on my PhD research in Professor Shinichi Hirai's Soft Robotics Laboratory at Ritsumeikan University. Professor Hirai is more than a normal supervisor, he is like an 'academic father' to me, who ultimately supports my ideas and is willing to discuss them with me, even late at night. Also, I would like to thank Dr. Dzung Dao, Prof. Sugiyama, Prof. Makikawa, and Prof. Kao for their support with experimental equipment, and significant advice that improved my research. Finally, I would like to thank my wife, Ngoc Anh, and my baby, Mit-chan, who give me joy and encouragement every day.

Kobe, October 2013 Van Ho

This book focuses on the slippage of soft fingertips in haptic sensing. Most of the chapters are based on the Ph.D. thesis of Dr. Ho, the coauthor of this book. He joined my laboratory in 2007 as a graduate student. I have been interested in human dexterity in object manipulation, and at that time, I wondered how human soft fingertips contributes to haptic sensing. Not only the pushing operation but also the sliding operation applies in the recognition of surface textures. The sliding operation implies that slippage between a human soft fingertip and a surface is essential in haptic sensing. At that time, mechanics of soft fingertips under pushing and rolling operations was studied extensively, but little attention was paid to mechanics of slippage. Dr. Ho and I discussed human haptic sensing and decided to formulate the mechanics of slippage for better understanding of human haptic sensing. Dr. Ho began to formulate the slippage of soft fingertips and developed the *beam bundle model*, which is the main concept of this book. From 2008 to 2011, he extensively studied the mechanics of localized slippage in haptic sensing and developed tactile sensing systems including a MEMS sensor and a fabric sensor to detect localized slippage. This was a very productive period.

Haptic sensing is important for dexterous object manipulation. Using our beam bundle model, we can formulate robotic object grasping and manipulation performed by soft fingertips with haptic sensation. Our beam bundle model applies to not only artificial fingertips but also human fingertips, and suggests that our model would be useful in the analysis of human haptic sensation. I hope that our research will be a valuable contribution to both robotic manipulation and human haptic sensing.

The authors would like to thank Dr. Thomas Ditzinger, Senior Editor of Springer, for his great effort on this book publication. In addition, I would like to thank my wife Chiaki and sons Tomoaki and Wataru for their cheerful encouragement.

Kusatsu, October 2013 Shinichi Hirai

Contents

Part II: Tactile Sensing of Localized Slippage

1

Introduction

TOUCH is a common but important action in everyday life. It permits us to accurately determine the surface properties and other properties of an object, including its weight and shape, which facilitates grasping tasks, as well as to determine many functions of the motor system. Until the 1970s, however, research on touching was limited to psychophysical studies. Given that the neural mechanisms underlying tactile sensation have been found to be critical to the success of adept manipulations [1], recent robotics research on dexterous manipulation has sought to imitate the natural touch mechanism, as well as the anatomy of human fingers, to optimize the ideal anthropomorphic artificial hand. Among the factors constituting an artificial hand, tactile perception is considered to be very important [2]. Without tactile feedback, failures in adept manipulation can occur, both to humans and robots. A sensory system must provide information about contact force, friction, and roughness, all of which are helpful for identifying objects. Moreover, during stable grasping, a sensory system must be able to recognize incipient and overt slips between the touching system and the object.

Therefore, of the various fingertip movements related to touch, slip perception by a fingertip plays an important role in dexterous manipulation. For example, to assess the texture of an object's surface, the fingertip needs to slide slightly over the surface to extract information about its roughness or friction (Fig. 1.1). The tendency of an object to slip through the fingertips during grasping, or incipient slip, is recognized as a crucial factor in stable manipulation of an object. However, while the latter studies have addressed all types of fingertip motion, such as pressing, rolling, and rubbing; the former studies focused primarily on pure pushing and rolling movements, while ignoring slip or slide with friction. Thus, it is necessary to obtain a fully described model of a sliding fingertip to answer the question *when exactly* and *how exactly* slippage occurs, so that such a model can be employed to assess slip perception in many applicable cases.

A.-V. Ho and S. Hirai, *Mechanics of Localized Slippage in Tactile Sensing*,
Springer Tracts in Advanced Robotics 99,
DOI: 10.1007/978-3-319-04123-0_1, © Springer International Publishing Switzerland 2014

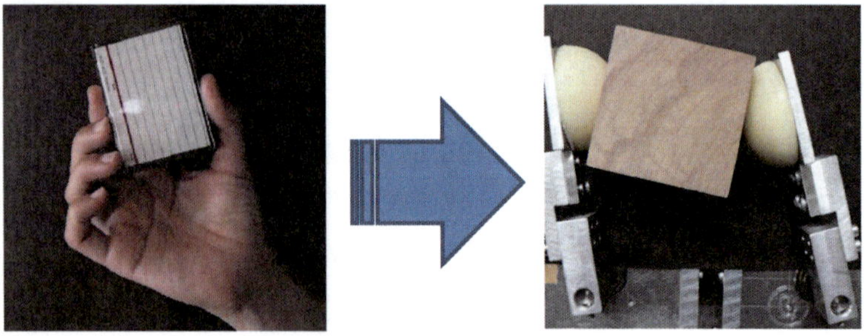

Fig. 1.1 Stable grasping by human and robotic hand, and the underline role of slippage. Mechanoreceptors of both the human and robotic hand continually check the slipping trend of a grasped object to output a suitable grip force to the motor system in order to maintain a stable grasp.

1.1 Related Work

1.1.1 Friction Model

One of the most challenging problems in slide movements is the appearance of friction. Due to difficulties in modeling friction and its unpredictability, friction is frequently neglected. Considerable theoretical and empirical research has been devoted to friction, including the Coulomb model and a complex model that combines Coulomb friction, viscous friction, and the Stribeck effect [3]. Basically, the frictional force between two contacting surfaces is a function of relative velocity V_r. However, the most challenging issue is its high nonlinearity at the point V_r=0. At this point, the amount of frictional force is unpredictable, and can vary between two thresholds [3]. This discontinuity at the origin is physically unrealistic and unacceptable for the purpose of simulation. Dahl [4] has hypothesized that frictional force originates from quasi-static contact bonds that are continuously formed and broken. Friction force in the Dahl model is a function of an internal state variable that can be interpreted as the displacement of the contact point. This leads to springlike behavior during stiction, with a lag in the change of friction force when the motion changes. The Dahl model has been widely used to improve performance while compensating for friction. By the introduction of damping, the Dahl model was developed into the LuGre model described in [5]. In addition, the Bristle model proposed by [6] was also recently shown to theoretically describe the complex behavior of friction. All of these models, however, are mostly specialized for rigid contact, and not for the contact of ductile, soft materials with a round shape, such as a soft fingertip.

Let us take a look at what happen on the contact surface during the sliding motion of a soft fingertip. Our objective soft fingertip is a semicylindrical soft fingertip, with radius r=10.0 mm and thickness q=3.0 mm. This soft fingertip

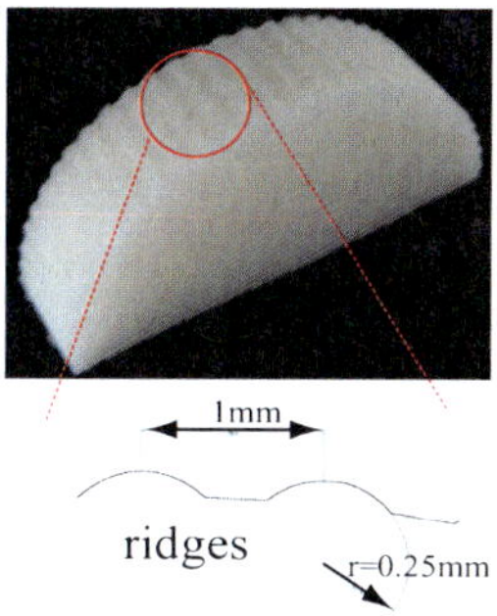

(a) Objective soft fingertip

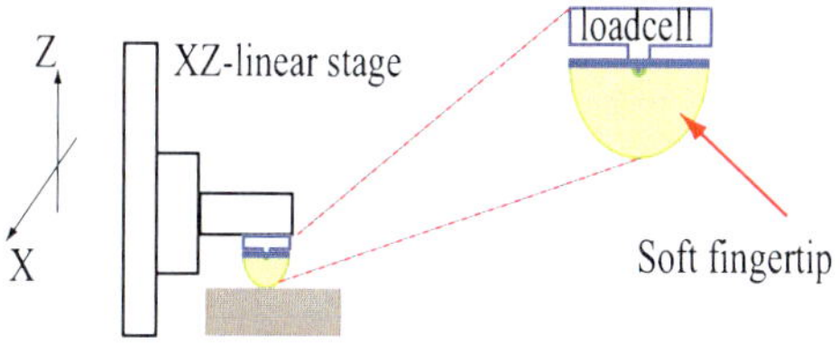

(b) Primary setup of experiment

Fig. 1.2 Experiment for assessing friction force acting on a sliding objective fingertip

is made from polyurethane rubber after an 8-hour curing phase at room temperature. The outer surface of the fingertip has many uniformly distributed ridges (Fig. 1.2(a)), each of which has the shape of a 90.0 deg arc of a circle with a diameter of 1.0 mm. The distance p between the centers of two adjacent arcs (or ridges) is about 1.0 mm. This distribution of ridges represents the epidermal ridges on a human fingertip. Figure 1.2(b) illustrates the setup of the experiment. The soft fingertip is attached to a 2-DOF (degree of freedom) XZ-motorized linear stage, allowing vertical and horizontal translation of the fingertip on a flat rigid plane. One 3-DOF load cell is attached to the fingertip to measure the total force (normal force and frictional force) acting on the fingertip during a sliding motion. The frictional force acting on the fingertip is recorded and plotted in Fig. 1.3. In this experiment, the fingertip was pushed for 1 mm at contact depth at a constant velocity of 1 mm/s and at a tilting angle of 0 deg. The frictional force can be split into two distinct phases: a stuck phase, in which the frictional force shows a steep increase; and a gross-sliding phase, in which the frictional force remains unchanged at a value of F_H(Fig. 1.3(a)). There is no sudden change in frictional force between the two phases at time t_0, *i.e.* the moment of the stick-slip transition. Moreover, when we assessed the hysteretic in a friction-velocity graph (Fig. 1.3(b)), we found that the frictional force acting on a sliding soft fingertip is quite similar to a Coulomb model with contact compliance.

Primarily, the frictional force acting on a soft fingertip does not differ greatly from other classical frictional models. Those models, however, cannot explain the partial movements on the contact surface of the fingertip theoretically, or determine in detail the timing and occurrence of sliding ([3] and [4]). These inabilities indicated the need to formulate a suitable contact model for soft fingertips that could approximate dynamically the multi-phases of contact.

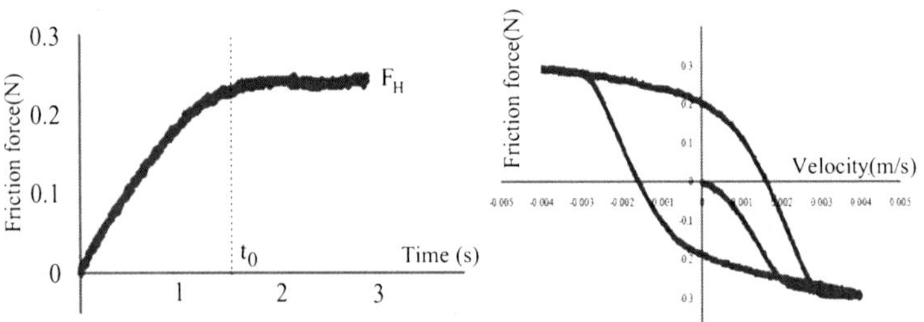

(a) Response of friction force (b) Friction force vs. velocity

Fig. 1.3 Responses of friction force during sliding motion. (a) Gradual change in friction when the fingertip switches from stick state to overt slip. (b) Hysteresis of friction during periodical movements.

1.1.2 Soft-Fingertip Model

Recently, several soft-fingertip models have been proposed that focus on analyzing the contact mechanics between various types of fingertips and objects. Xydas and Kao [7] proposed a model for a hemispherical soft fingertip using finite element (FE) analysis utilizing the nonlinear characteristic of a material. Maeno *et al.* [8] employed a complex model of a soft fingertip also using FE methods to assess the quasi-static results of stress-strain relations during contact with objects. While FE analysis allows us to derive the exact stress distributions within a fingertip and on a contact surface, it cannot bring dynamic change during contact, and in addition, its implementation is expensive. There are, in addition, analytical ways to represent deformation of a contacting soft fingertip by utilizing elastic potential energy. Nguyen *et al.* [9] derived an elastic force model for potential energy by employing virtual linear springs that were arranged, for simplicity, along the radial directions of a hemispherical fingertip. Inoue and Hirai [10], conversely, aligned virtual springs normal to the fingertips, and proposed the theory of local minimum of elastic energy (LMEE) which was shown to be efficient in controlling manipulation. Authors in [11] investigated soft-finger contact by demonstrating contact with a set of linear springs based on screw theory, revealed the rotational effect, and identified the stiffness properties of the contact. However, those models mainly focus on pushing and pure rolling action, and none of them addresses the problem of sliding motion, especially with the appearance of friction.

The difficulties in modeling the sliding motion of fingertips stem from the compliance of fingertips, their partial movements on contact surfaces, and friction force/moment. Kao and Cutkosky [12] have proposed a method that combines compliance and friction on a limited surface to compute the relative sliding motion between a grasped object and soft fingers. In their paper, they showed concrete results in modeling contact, and in approximate gross-slip motion planning. Authors in [13] introduced an algorithm for automatic stable grasping of polygonal objects with two fingers and point contact with friction. Related research can be found in [14] and [15]. Recently, authors in [16] investigated the sliding of a tiny hemispherical rubber nub glued to the free end of a loading beam. The frictional characteristics were modeled utilizing Galerkin's approximation of an Euler-Bernoulli beam and the experimental results showed the correctness of this model. However, most research has consisted only of quasi-static analysis and gross sliding, with little consideration being paid to the stick-slip phases (how and when it happens), or partial slippage on the contact surface of a soft fingertip. Experimental results [18] have concluded that the shape of the fingertip causes complications in sliding motion, including partial localized movements on the contact surface, especially during the stick-slip phase.

1.1.3 Slip Detection Attempts

Recently, in conjunction with soft fingertip modeling, there have been many studies concentrating on the tactile texture perceptions of sensory fingertips with the aim of increasing efficiency during object manipulation processes, especially with respect to the detection of slippage.

Among the sensors utilized was a skin acceleration sensor, which consisted of a thin outer skin of rubber bonded to a hemicylindrical core of foam rubber [17]. Two interferometers were attached to this skin that responded to its vibration. Vibration is considered to be an indicator of the status of contact, such as slipping or sliding, and experimental evidence indicated that, by using these sensors, incipient slippage could be detected efficiently. Multi-element stress rate sensors, consisting of strips of piezoelectric film, a silicon rubber skin and a foam filling, have also been used to detect localized incipient slipping [18]. Four piezoelectric film strips were implanted under the skin and stretched across the entire width of the fingertip. Using this design, incipient slippage could be detected and curvatures localized. Similarly, PVDF (polyvinylidene fluoride) film sheets, which provide dynamic stress rate characteristics, have been used to partly construct artificial finger skin for realizing contact [8]. This skin, which imitates the characteristics of human fingers, consisted of nine ridges; each of which included two PVDF films perpendicular to each other and enabling detection of the vibration of that ridge. A multi-layer artificial neural network (ANN) was used to assess the realization of incipient slips. In designing a texture sensor that imitates a human finger, authors in [19] described a multi-layered structure of soft material, mechanoreceptor, and ridges. Several experiments were performed to assess the ability of this system to quantitatively perceive the texture of various materials. These integrated systems, however, were found to be difficult to design and implement. Although the sensors were implanted near the outer skin, allowing them to respond sensitively to states of contact, this caused complications in fabrication and data acquisition due to their volume and output wiring. It was therefore necessary to develop a miniaturized sensory device that is multi-functional and easy to integrate into tactile systems.

Recently, interest has increased in fabricating micro-electro mechanical (MEMS) sensors artificial neural network, which can take advantage of the highly integrated functions and packaging of MEMS technology. Engel *et al.* [20] developed an integrated flexible tactile sensing skin that was sensitive not only to common surface characteristics such as frictional force and roughness, but also to thermal conductivity, hardness, and temperature. To the best of our knowledge, however, these sensors have not been utilized in robotic manipulation, including incipient slip detection. Beccai *et al.* [21] fabricated a tactile sensing skin using miniaturized silicon-based sensors that could detect three components of external force. A joystick-like mesa was attached to a sensor base to transfer external force. This was later embedded under a polyurethane skin to form a soft, compliant, tactile micro sensor. This system was shown experimentally to be capable of detecting incipient slip transition because the sensor was embedded just under the elastomeric skin, which for its part is highly sensitive to the disturbance and danger of fragileness. The mesa is easily broken under high loads when the tactile systems grasp a sufficiently heavy object. A similar design for a tactile sensor, proposed by [22], showed the ability to sense the structure of a fabric or to differentiate papers by using supervised classification methods. In addition, several three-axial sensors in [23]-[24] haven been also implemented, but none of them have addressed applications in artificial tactile sensing experiments.

1.2 Our Approach

In this book, we concentrate on comprehensively describing the dynamic sliding motion of soft fingertips, especially stick-slip transition, which is considered crucial in stable grasping/manipulation. The sliding of soft fingertips, unlike that of rigid ones, is more complicated as a result of its ductility, nonuniform shape, and friction. We assume that the volume of the fingertip is filled with elastic cantilevers that are compressible and bendable when the fingertip slides. We also assume that contact surface is meshed so that position of one contact node is coincident to that of one corresponding beam. This *hybrid* paradigm helps in conducting dynamic simulation as well as reducing simulation cost remarkably. We call this model the *Beam Bundle Model* (BBM).

Our objective is to assess *how* and *when* slippage occurs from the rest state of a sliding soft fingertip. Modeling the fingertip using the hybrid model enables us to elaborate the micro slips of contacting nodes, telling us which parts of the contact surface are still sticking, which parts have already given way during the stick-to-slip transition. By looking at the ratio of slipped zones over the contact surface, we are able to *track* the slipping trend, which is, as mentioned above, crucial to assessing overt slip.

In our first attempt, we construct a 2-D model of a sliding soft fingertip that has uniform depth, such as a cylindrical finger. Through simulation trials of stick-to-slip transition, we are able to obtain the friction response, and its relation to an object's velocity or displacement. We also present the idea of localized displacement phenomenon (LDP), and a method to detect the *incipient* slip of the sliding fingertip. The simulated results are verified by fine setup experimental results.

In our second attempt, we extend the model into the 3-D space of a hemispherical sliding soft fingertip, which is the BMM. This model can be generalized as a platform to model various shaped soft fingertips provided the outer shape of the fingertip is known. In this simulation, we also explore the idea of LDP in a 3-D model. We then examine the effect of friction torque, which is unique for compliant contact.

In the third attempt, we make use of the BMM to model several soft tactile systems, and also use the LDP theory to assess slip perception for each system. First, we fabricate a tactile soft fingertip embedded with a micro force/moment sensor (MFMS) and propose an efficient way to detect incipient slip using the LDP theory. Second, a robotic skin sensor made from pressure-sensitive electro-conductive yarns is constructed as a collaborative study utilizing the idea of LDP. We then propose a method to detect fast slippage on the surface of this sensor, especially human rubbing action, by employing a discrete wavelet transform (DWT), and endow it with texture discrimination ability.

In the forth attempt we propose a promising method to detect slippage of a grasped object by using tactile-arrayed sensors attached to a gripper. By utilizing the LDP idea and image processing, for the first time we are able to assess the tangential slip of the object, even although the tactile sensor only can obtain information on normal stress.

1.3 Organization of the Book

The rest of this book is organized as follows. Chapter 2 and Chapter 3 describe the attempts made to model and simulate a sliding soft fingertip, focusing on stick-to-slip transition, with Chapter 2 describing how the 2-D model introduced in Chapter 1 is extended to a 3-D simulation (BMM). Here, we characterize the friction response on a contact surface, and the micro movements of contacting points during stick-to-slip transition, then propose the idea of LDP in detecting the slip action of soft objects.

In the second part, Chapters 5 to 7 discuss the application of LDP in soft tactile systems. Chapter 5 describes our collaboration with the Micro Nano Integrated Devices Lab (Risumeikan University) to develop a soft tactile fingertip embedded with a MFMS for use in detection contact states, especially slip detection. In Chapter 6, we detail our research to endow a low-profile fabric sensor with intelligence of slip perception and texture discrimination by utilizing sling motion and the LDP idea. In Chapter 7, we introduce an idea *From tactile to image processing*, which proposes a soft-pressure array sensor using various perceptions from localization, contact shape recognition, to slip detection that was inspired by LDP.

Finally, the findings of this book are summarized in Chapter 8 with a discussion and conclusions, as well as plans for future work.

Part I

Mechanical Modeling of Localized Slippage

2

Two Dimensional Beam Bundle Model of a Frictional Sliding Soft Fingertip

We propose a dynamic model of a soft fingertip to investigate its sliding motion on a plane with friction. The fingertip is comprised virtually of a finite number of elastic compressible and bendable beams whose free ends act as contact points. Moreover, we utilize virtual linkage spring-damper elements between the contact points to represent interactions between neighboring beams on the contact surface. By introducing Coulomb's law into each contact point, we are able to capture the frictional characteristic during sliding motions of the fingertip, especially stick-slip motions. During analysis, we have found a phenomenon called *Localized Displacement* which takes an important role in assessing slip perception of soft object. This phenomenon is then verified by our fine experiential setup.

2.1 Two-Dimensional Modeling of Sliding Soft Fingertip

2.1.1 Beam Model of a Soft Fingertip

Previous research in [10] proposed that a soft fingertip model comprised of an infinite number of vertical elastic virtual springs that could be used to investigate the deformation of the fingertip during pushing or rolling motion on an object. This model, however, was not sufficient to demonstrate sliding motion with the appearance of frictional force. Therefore, instead of virtual springs, we proposed a new model, in which each soft fingertip was composed of a finite number of *virtual elastic beams* that were compressible, tensile, and bendable (Fig. 2.1). This new model, thus, can represent the diverse deformations of a soft fingertip during a sliding motion, in which a soft fingertip is pushed and slid at the same time. These beams are fixed on the equatorial surface of the fingertip, with their free ends on the outer surface of the fingertip. Each beam has a uniform cross sectional area, of width p and thickness q;

A.-V. Ho and S. Hirai, *Mechanics of Localized Slippage in Tactile Sensing*, Springer Tracts in Advanced Robotics 99,
DOI: 10.1007/978-3-319-04123-0_2, © Springer International Publishing Switzerland 2014

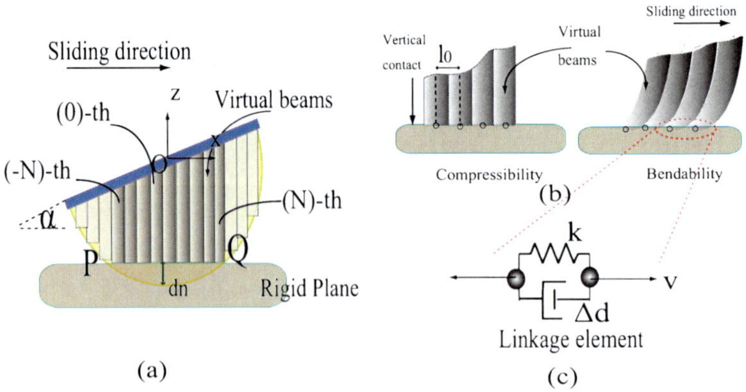

Fig. 2.1 Proposed model of a sliding soft fingertip (a) Model filled by virtual beams. (b) Compressibility and bendability of beams. (b) Model of a contacting point at one beam's free end.

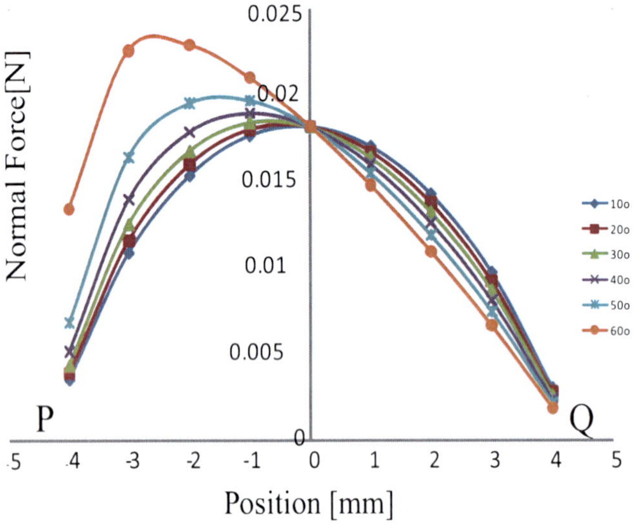

Fig. 2.2 Distributions of normal force over contact line when titling angle changes

while the lengths of the beams differ depending on their coordinates within the fingertip.

Conventionally, previous theoretical research eliminated the role of frictional force in dynamic modeling. However, frictional force plays a crucial role in the stabilizing of objects grasped by robotic hands (see [10],[9]). Our advanced model, which divides the soft fingertip into many beams, splits patch contact between the fingertip and the object into point-to-surface contacts between the free ends of the beams and the surface of the objects. This allows Coulomb's friction law (see [25]) to be applied to each free end of the beam on the contact surface. As a result, by observing the dynamic movements of all free ends of the beams, we could assess the transient responses, especially the incipient slip, during the sliding motion of the fingertip.

Using r as the radius of the soft fingertip, we first formulated the distribution of forces inside the soft fingertip when it is pushed perpendicular to the plane with contact depth d_n. In this case, all beams are pushed vertically, with different normal deformations. The normal force F_i^n, acting on one free end of one beam whose coordinates are determined by x, and titling angle α (Fig. 2.1(a)), could be calculated based on the equation

$$F_i^n = E\frac{\pi pq}{x\tan\alpha + \sqrt{r^2 - x^2}}(d_n - (r - \sqrt{r^2 - x^2})), \tag{2.1}$$

where E denotes the Young modulus of the beam. Fig. 2.2 illustrates calculated normal force distribution along contact patch when the fingertip is pushed with the same contact depth, and tilted with different angles. We can observe that, the distributions are not uniform. The highest value shifts away from center point when titling angle α varies; and the total normal force increases with the larger contact force. This suits to conclusions generated in previous research [10].

For the sake of simplicity, we made three assumptions

1. When a beam is bent, its deformation is significant only at the free end.
2. Interactions between continuous beams only occur between their free ends on the contact surface.
3. Only beams whose free ends are acting on the contact surface are considered (dark colored beams in Fig. 2.1(a)). Beams positioned outside the contact surface are deemed irrelevant to the sliding motion of the fingertip (light colored beams in Fig. 2.1(a)).

Based on these assumptions, we determine the interactions among virtual beams, as well as between the beams and the contacting plane, to demonstrate the more precise behavior of the fingertip during its manipulating task. The contact element of each beam can be reduced to a simplified point contact model. Between two neighboring contacting points, there is a virtual *linkage* element, or Voigt model, consisting of an elastic element (spring) and a viscous element (damper) connected in parallel (Fig. 2.1(c)). The *linkage* spring with stiffness k represents the interaction between beams, as well as the elasticity of the point of contact; while the damper with damping coefficient Δd represents the viscosity on the contact surface. Each contacting point is assigned a point mass Δm. Friction acting on a contacting point with a coefficient of friction μ is dependent on the velocity of the contacting point.

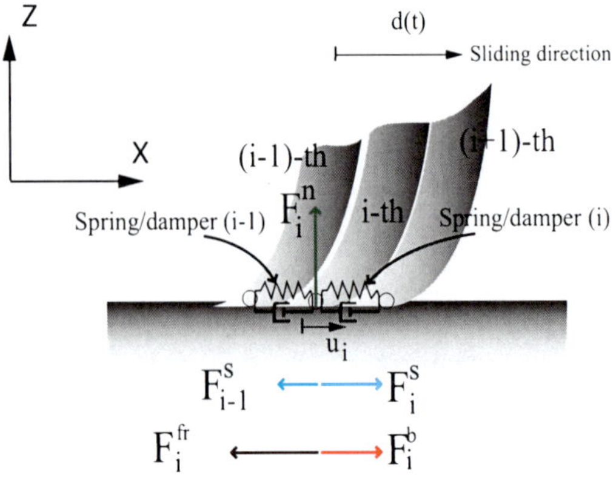

Fig. 2.3 Analysis of forces acting on the i-beam

2.1.2 Physical Parameters of Beams

We first measured Young's modulus E of the soft fingertip used in the simulation by performing a compression test on polyurethane gel and implementing a linear approximation, finding that E=0.606 MPa. Number of beams is dependent on the contact depth d_n and beam's width p. The number of contacting points is $2N+1$, numbered from $(-N)$-th to N-th along the direction of sliding motion (Fig. 2.1(a)). We set l_0 as the distance between the neutral axes of two adjacent beams (Fig. 2.1(b)). The bending stiffness of the i-th beam was then calculated based on its parameters, as well as on the equation

$$\Delta b_i = \frac{3EI}{{l_i}^3} = \frac{Eqp^3}{4{l_i}^3}, \tag{2.2}$$

where l_i be the natural length, I be inertia moment of the i-th beam. The stiffness of each linkage spring is specified by

$$k = \frac{Es_0}{l_0} = \frac{E\pi {d_{r_0}}^2}{l_0}, \tag{2.3}$$

where d_{r_0} is the radius of the cross sectional area of the linkage spring. Equations 2.2, and 2.3 can be found in [28]. While it was relatively easy to determine the parameters of the spring, it was more difficult to analyze the damping coefficient. Basically, Δd can be determined by following equation

$$\Delta d = 2\zeta\sqrt{k\Delta m}, \tag{2.4}$$

where Δm is mass at contact point, and ζ is damping ratio.

The plane was fixed; while the fingertip was given vertical and horizontal constraints. By pushing the fingertip with a constant contact depth of d_n, each beam will receive a specified normal compression, and, based on Eq. 2.1, the value of the normal force F_i^n acting on it can be calculated. Subsequently, at one specific contact depth, the fingertip is given a displacement $d(t)$ along the positive direction ($\dot{d}(t) > 0$). Using these constraints, a model of a sliding fingertip on a plane could be assessed. The motion of slide of the fingertip is categorized into three phases stick-slip phase, gross slide phase, and reverse of slide. To elaborate the various phases of a sliding fingertip, we attempted to build equations of motion for each phase.

Initial Stick-to-Slip Phase

Let us detach the i-th beam to assess the external forces acting on it at time t. Fig. 2.3 shows the detached i-th beam and interactions with its two neighboring $(i-1)$-th, and i-th linkage springs. The normal force F_i^n is caused by normal deformation of the fingertip, or beam, and calculated by equation (3.1). Using u_i, u_{i-1}, and u_{i+1} be the generalized displacements of the free ends (or contacting points) of the i-th, $(i-1)$-th, and $(i+1)$-th beams, respectively. Two interaction forces F_i^s, and F_{i-1}^s were caused by deformations of the $(i-1)$-th and i-th linkage spring/damper elements when the free end of the i-th beam starts moving; and are described as follows:

$$F_i^s = k(u_{i+1} - u_i) + \Delta d(\dot{u}_{i+1} - \dot{u}_i), \tag{2.5}$$

and

$$F_{i-1}^s = k(u_i - u_{i-1}) + \Delta d(\dot{u}_i - \dot{u}_{i-1}). \tag{2.6}$$

In the initial stick-to-slip phase, the generalized displacement u_i of the contacting point, and $d(t)$ of the fixed end of each beam starts at zero. Therefore, the bending strain $s_i(t)$ can be calculated as

$$s_i(t) = d(t) - u_i, \tag{2.7}$$

yielding to bending force acting on the contacting point

$$F_i^b = \Delta b_i s_i(t) = \Delta b_i(d(t) - u_i). \tag{2.8}$$

Moreover, tangential frictional forces are found to act on the free end of the beam F_i^{fr}. Consequently, the equation of motion the i-th beam is formulated as

$$F_i^b + F_i^s - F_{i-1}^s - F_i^{fr} = \Gamma_i, \tag{2.9}$$

or

$$\begin{aligned}&\Delta b_i(d(t) - u_i) + k(-2u_i + u_{i+1} + u_{i-1}) + \\ &+\Delta d(-2\dot{u}_i + \dot{u}_{i+1} + \dot{u}_{i-1}) - F_i^{fr} = \Gamma_i.\end{aligned}$$

Note that Γ_i depending on the states of contact of the free end of each beam. During the sticking period, $\Gamma_i = 0$, and friction could be calculated based on the following equation

$$F_i^{fr} = \Delta b_i d(t) + (k(u_{i+1} + u_{i-1}) + \Delta d(\dot{u}_{i+1} + \dot{u}_{i-1}). \qquad (2.10)$$

When F_i^{fr} reaches $F_i^n \mu$, the free end of beam starts to move based on Coulomb's law, $\Gamma_i = \Delta m \ddot{u}_i$. We assume the of static friction coefficient μ has the same value for all contacting points.

For other beams, the scheme of acting forces are similar. Let $\mathbf{u} = [u_{-N}, .., u_i, .., u_N]^t$ be a collective vector of the generalized displacement of all beams, $\mathbf{B} = [k^b_{-N}, .., k^b_i, .., k^b_N]^t$ be vector of bending stiffness of beams, the motion equations for all the beams, *i.e.* a soft fingertip, during sliding motion could be summarized as followed

$$\mathbf{Ku} + \mathbf{D\dot{u}} - \mathbf{F_{fr}} + \mathbf{B}d(t) = \Delta m \ddot{\mathbf{u}}, \qquad (2.11)$$

where $\mathbf{F_{fr}}=[F^{fr}_{-N}, .., F^{fr}_i, .., F^{fr}_N]^t$, and

$$\mathbf{K} = \begin{pmatrix} -\Delta b_{-N} - k & k & & \\ k & -\Delta b_{-N+1} - 2k & k & \\ & \vdots & \vdots & \vdots \\ & & k & -\Delta b_N - k \end{pmatrix},$$

$$\mathbf{D} = \Delta d \begin{pmatrix} -1 & 1 & & \\ 1 & -2 & 1 & \\ & \vdots & \vdots & \vdots \\ & & 1 & -1 \end{pmatrix}$$

are referred as contact stiffness matrix and contact damping matrix, respectively. We use a numerical method to assess the movements. During the simulation time, the value of the frictional force acting on each beam's free end is calculated and used to determine if the free end sticks or slips. The overt slip of a soft fingertip on the contact surface will occur when all the free ends slide.

Gross Sliding

After the last contacting point slides in the initial stick-to-slip phase at moment t_0, the contacting line between the fingertip and plane will slide stably, simultaneously with the movement of the fingertip. In detail, the bending strain $s_{0i} = d(t_0) - u_{0i}$, and the distance between two adjacent contacting points $|u_{0i} - u_{0i-1}|$, *i.e.* deformation of one linkage element, will maintain the last state of the stick-to-slip phase; while $\dot{u}_i$ equals to $\dot{d(t)}$. During this period, adjacent contacting points stop interacting, and linkage springs/dampers no

Table 2.1 Parameter Values Used in Simulation

Parameter	Value	Unit
r	10	mm
d_n	1	mm
p	1	mm
q	3	mm
d_{r_0}	1	mm
2N+1	9	beams
k	650	N/m
Δd	1.6	Ns/m
δd	0.9	Ns/m
Δm	0.002	kg
μ	0.6	

longer contribute to friction. Instead, the frictional force of each contacting point will be a combination of Coulomb friction and viscosity friction (with viscous friction coefficient δd), which will be dependent to velocity $\dot{u}_i$ as:

$$F_i^{fr} = F_i^n \mu + \Delta d \dot{u}_i = F_i^n \mu + \delta d \dot{d}(t). \tag{2.12}$$

Stick-Slip Phases at Changes in Direction of Slide

When velocity of the fingertip $\dot{d}(t)$ goes to zero, the state of frictional force changes from sliding to sticking. When $\dot{d}(t)$=0, the friction force switches from Eq. 2.12 to Eq. 2.10. However, due to the initial deformations of linkage springs/dampers calculated during the gross-slide phase, Eq. 2.10 is modified as

$$F_i^{fr} = \Delta b_i s_{0i} - k(-2u_{0i} + u_{0i+1} + u_{0i-1}). \tag{2.13}$$

When the fingertip starts to slide in the reverse direction ($\dot{d}(t) < 0$), the process from sticking to slipping occurs similar to the initial stick-to-slip phase. Equations from (2.5) to (2.11) are reused with the addition of the initial deformation of the linkage elements, and adjustment of the sign due to the change in direction.

2.2 Numerical Simulation

To assess the behavior of frictional force, several simulations were conducted that are similar to [16]. Parameter values used for the simulation are summarized in Table 2.1. The value of the damping coefficient Δd was chosen to give a critically damped behavior $\zeta = 0.8$ with a mass Δm=0.002 kg.

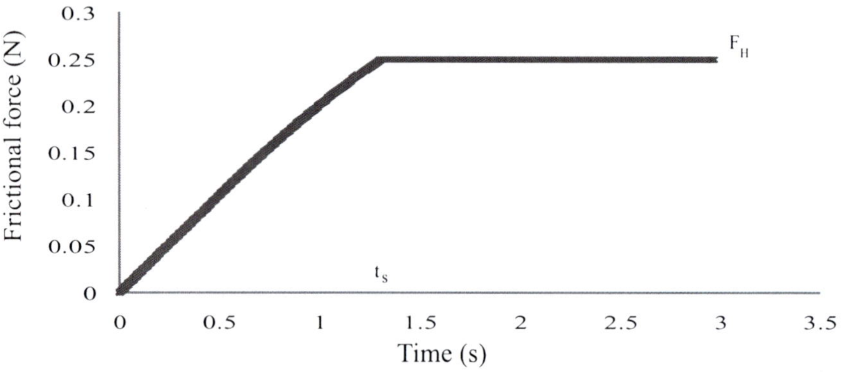

Fig. 2.4 Friction at constant velocity

Coefficient of static friction is given $\mu = 0.6$, and viscous friction coefficient $\delta d = 0.9$ Ns/m are considered. We used 4-th order Runge-Kutta ODE integrator with a constant time-step in this numerical simulation. We will first assess the stick-to-slip phase to determine how and when the fingertip slides and to acknowledge the incipient slip which is important in stable grasping manipulations. Subsequently, we show a periodic movement of the fingertip to further determine the characteristics of frictional force during sliding.

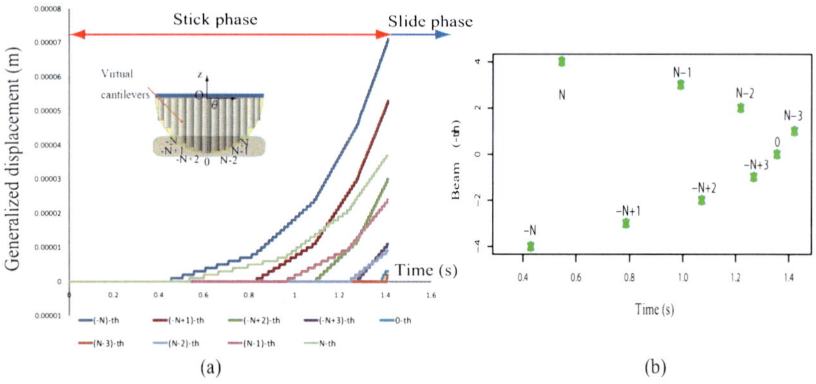

Fig. 2.5 a) Localized movements of contacting points during initial stick-to-slip phase. The inset picture illustrates numbered beams. b) Orders over time of first slides of contact points during stick-slip phase.

2.2.1 *Stick-to-Slip Phase, and the Localized Displacement Phenomenon*

Fig. 2.4 shows the response of total frictional force acting on a contact line during a stick-to-slip process, when the fingertip has the same contact depth and velocity as those in the experiment described in Section 2.1. This process obeys the classical Coulomb's law, in which the frictional force increases during the stuck phase and changes slightly to the sliding phase without a sudden nonlinear step. This phenomenon occurs only with ductile material like a soft fingertip; but is unrealistic for solid objects when the body accelerates even though the external forces on the body are less than the threshold of frictional force (see [6]).

As a result, the response of this simulation is similar to that of the experimental frictional force shown in Fig. 1.3(a), including the kinetic value (F_H) and the moment of slip (t_0). Values of (F_H, t_0) in Fig. 1.3(a) (see Chapter I) were slightly higher than those in Fig. 2.4 ((0.27 N, 1.49 s) and (0.25 N, 1.42 s), respectively). Fig. 2.5 shows the movements of contacting points during this phase. Micro slips of contacting points are recorded with no obvious vibration, thanks to the critically damped criteria that we chose. As a result, there are *partial movements* of contacting points during the stuck phase on distributed areas of the contact surface. We call this as *Localized Displacement Phenomenon* (LDP), which dominates transition from stick to slip of a contacting soft fingertip.

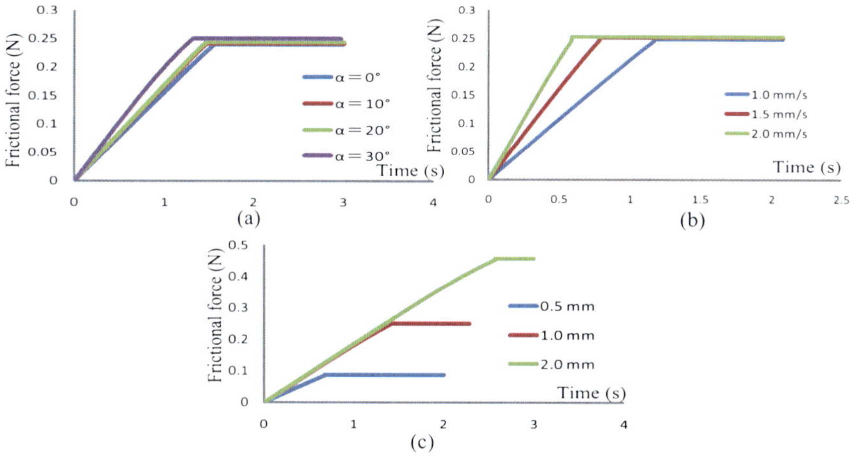

Fig. 2.6 Responses of frictional force during stick-slip phase when a) Tilting angle α changes. b) Relative velocity V_r changes. c) Contact depth d_n changes.

During this phenomenon, the order of first slide of one contacting point was different than that of others. If the direction of slide is from left to right (Fig. 2.5(a), inset), there would be two trends of slide propagation. The first one is from the most outer contacting point (for example, the $(-N)$-th and the N-th contacting points) to the inner ones ($(-N+1)$-th, $(N$-1)-th). The other one of slide propagation is from left to right $(-N+1)$-th, $(-N+2)$-th, *etc.*). Finally, the last contacting point to move before gross slide is about in the middle of the second half of the contact line (Fig. 2.5(b)). As a result, *Localized Displacement Phenomenon* is found in the simulation, indicating that during the transition from stick to slip states there *exist* moved points on the contact surface, and the gross slippage of the soft fingertip *only* happens when all the points on the surface give ways. This phenomenon is considered important in assessing when and how the slippage occurs on the contacting surface of soft fingertips with object, during stick-to-slip transition.

We varied some conditions of a sliding fingertip to assess more about this stick-slip phase. In Fig. 2.6(a) are plots of frictional force when the fingertip's titling angle α increases from 0 to 30 deg. Values of F_H and t_0 are slightly higher and smaller, respectively, in companion with increase of α. As a result, despite of higher value of total normal force when α increases (as shown in Fig. 2.2) the stick-slip transition happens earlier. Changes of (F_H, t_s) also could be observed clearly when the velocity changes (Fig. 2.6(b)), or the contact depth changes (Fig. 2.6(a)), both when $\alpha = 10$ deg.

2.2.2 Reverse Phase

In this phase, the velocity of fingertip $\dot{d}(t)$ changes from positive to negative sign (Fig. 2.7(a)). Frictional force switches from slide to stuck phase when $\dot{d}(t)$=0, as determined from equations (2.12), and (2.13), respectively. When the sign of the velocity $\dot{d}(t)$ changes, each beam tends to slide ahead (*i.e.* in the positive direction) due to its initial bending strain s_{0i}. Therefore, when the fingertip changes direction, the bending strain will decrease from s_{0i} to zero, causing the frictional force to decrease, while maintaining the same sign.

Subsequently, however, the direction of frictional force will change, going into the stick-to-slip phase, as described above. During this transition, the frictional force is only recorded with some unremarkable disturbances, as shown in Fig. 2.7(a). This reverse phase will take longer than the initial stick-to-slip phase. As a result, this model can more closely reflect reality, in contrast to earlier models of sliding soft fingertips that was described in [16], in which this transition caused friction to change randomly, causing complications in both simulation and control.

2.2.3 Periodical Movement

To further assess the behavior of frictional force, we conducted a simulation in which the fingertip moves periodically with velocity $\dot{d}(t)$, as plotted in

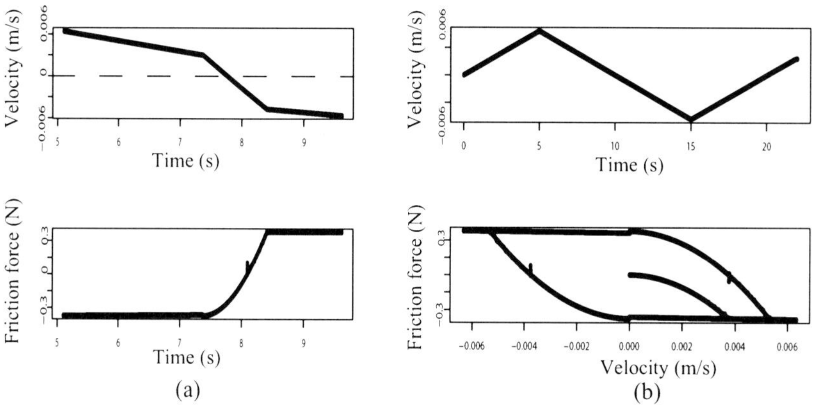

Fig. 2.7 Responses of frictional force when the relative velocity changes a) Reverse of velocity. b) Periodical change of velocity.

Fig. 2.7(b). The graph of frictional force-velocity shows a similar tendency to that in Fig. 1.3(b). As described in Section 2.1, hysteresis was included in this plot. Therefore, by using this model, this important characteristic of frictional force can be described theoretically. As a result, this relationship between frictional force and velocity can be used to predict the velocity when the fingertip slides. Both graphs, however, show that the initial stick-slip phase can be distinguished from other stick-slip phases, such as reverse phase. This was noted in the above prediction. Moreover, when we change the rates of velocity changes, we find that the width of the hysteresis of friction-velocity increases, while that of friction-displacement decreased horizontally with increase in frequency (Fig. 2.8(a)), as previously described in [16]. By examining the friction-displacement relation, we also observe a steep slop in the region where sliding changes its direction, as plotted in Fig. 2.8(a). Authors in [5] describe this as "spring -like" sticking behavior. When the cycle time of periodical movement increases, square enclosed by the hysteresis of friction-displacement increases as well.

The changes of width of these hysteresis are also observed when the contact depth changed. In this case, both hystereses of friction-velocity and friction-displacement got bigger vertically due to a change in frictional force that occurred when the contact depth varied (Fig. 2.8(b)). In the other investigation, the hysteresis of friction-velocity seemingly rotates around original point 0 when the tilting angle α changes from 0 deg to 30 deg (Fig. 2.8(c)). The difference can be observed clearly with large value of α.

In conclusion, by conducting simulation of sliding motion of the fingertip with proposed model, we can describe multiple states of friction. It depends remarkably on conditions of sliding of the fingertip, such as contact depth, tilting angle, and relative velocity. Especially, for the first time, the *Localized*

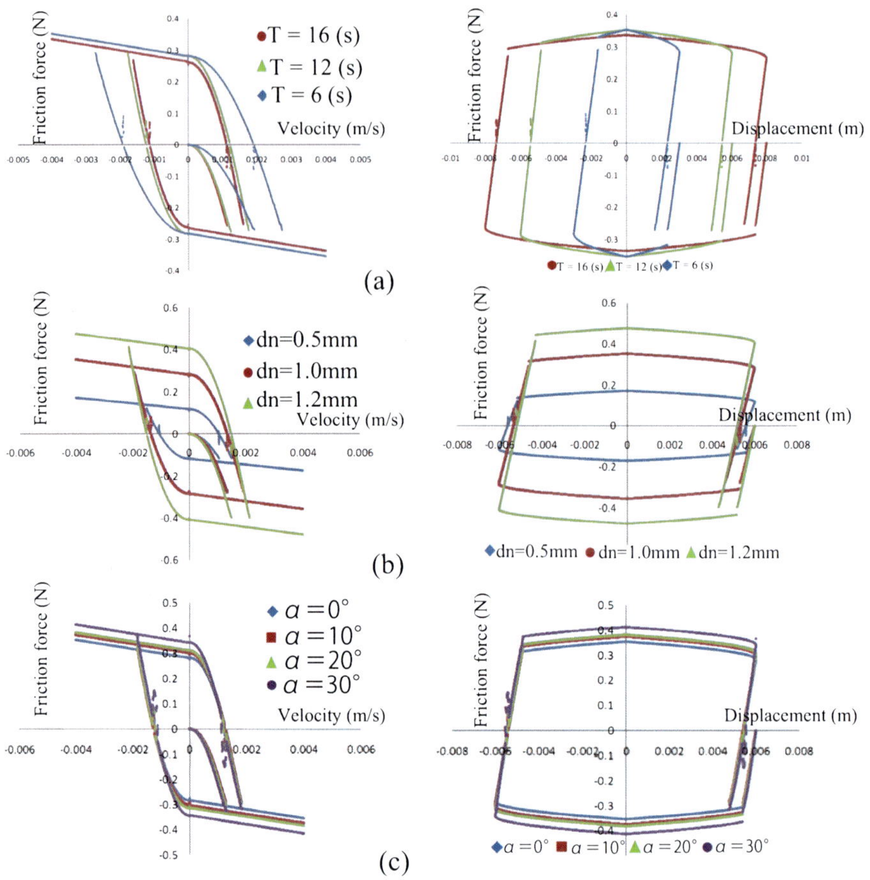

Fig. 2.8 Graphes of friction-velocity and friction-displacement relations during periodical movements of the fingertip when a) Cycle time T changes. b) The contact depth d_n changes. c) The tilting angle α changes.

Displacement Phenomenon is found in simulation during stick to slip transition. By discovering this phenomenon, we are able to understand how and when the slippage occurs, as well as predict the micro slippage on the contact surface. This promises to application of slip detection of tactile sensing systems.

2.3 Experimental Validation

Configuration of the experiment differed slightly from that described in Fig. 1.2 (see Chapter 1). The rigid plane was replaced by a transparent rigid plane, allowing a high speed camera (MV2-D1280-640 CMOS camera, Photonfocus) to take successive images of the contact surface during sliding motion of the fingertip at a rate of 500 fps (frames per second). This setup will permit a closer look at tiny movements of contacting points. To help the camera more easily identify ridges on the surface of the soft fingertip, we coated the ridges with a very thin layer of black paint.

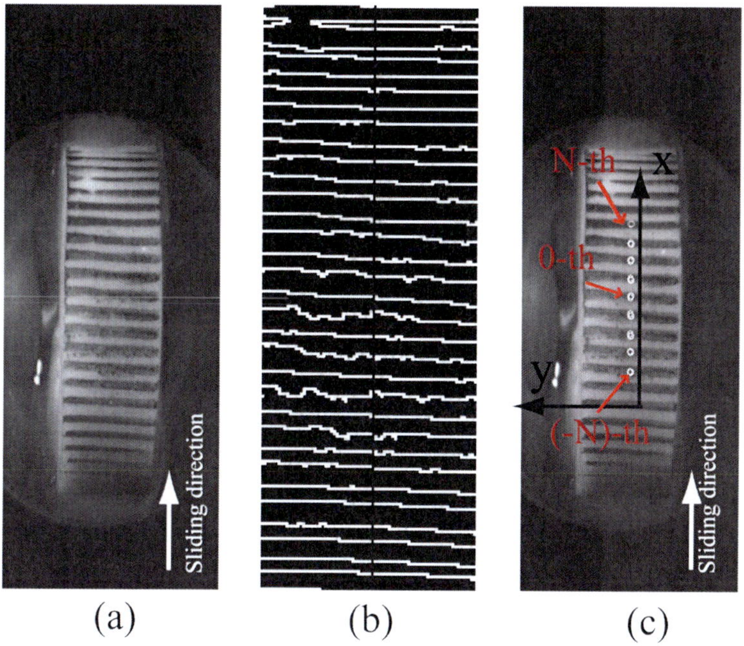

Fig. 2.9 Images for processing (a) Gray-scale image. (b) Binary image with extracted edges. (c) Image with tracked points.

2.3.1 Verification of the Localized Displacement Phenomenon

This experiment was designed to compare the localized movements of contacting points in stick-to-slip phase in simulation and reality. Fig. 2.9(a) shows a grayscale image of ridges on a fingertip's skin with the contact surface. By conducting image processing using OpenCV, we obtained binary images with coordinates of contours (Fig. 2.9(b)). Subsequently, we observed the movements of the center points of contacting ridges on the contact surface (Fig. 2.9(c)). These center points represent contacting points in the model. The i-th ridge was extracted into two continuous vertical edges by using edge-emphasizing filter. Central points' coordinates of these edges, say (x_i,y_i) and (x_{i+1},y_{i+1}), are perceived afterward. Ultimately, tracked point of the i-th

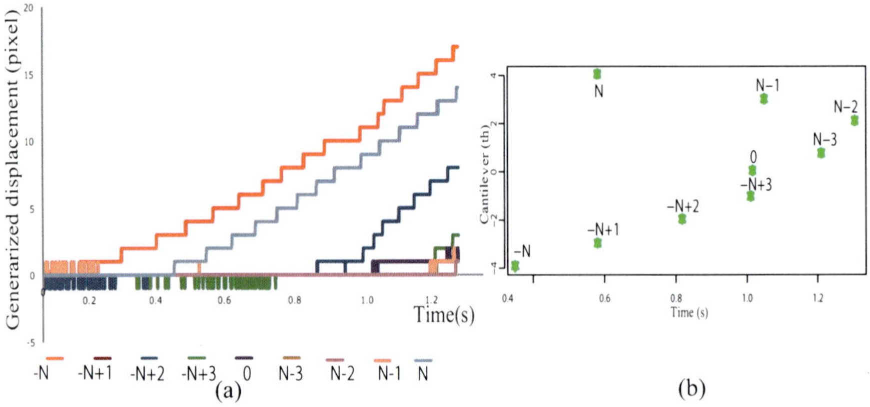

Fig. 2.10 a) Localized movements of contacting points during initial stick-to-slip phase. b) Orders over time of first slides of contact points during stick-slip phase.

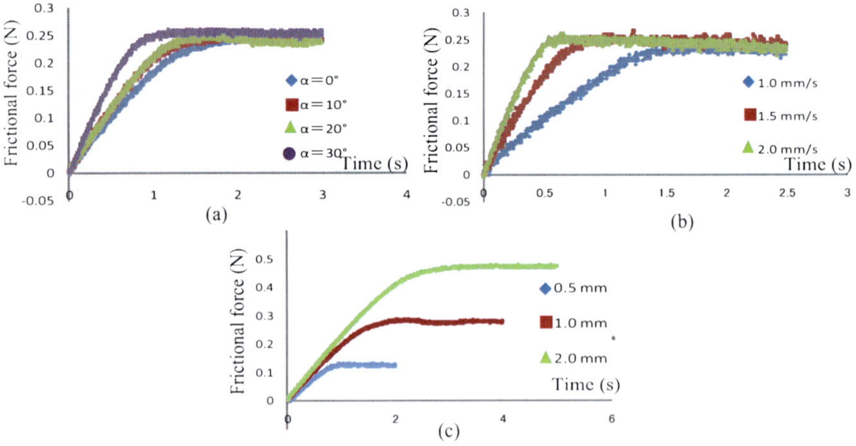

Fig. 2.11 Experimental responses of frictional force during stick-slip phase when a) Tilting angle α changes. b) Relative velocity V_r changes. c) Contact depth d_n changes.

ridge is defined as $1/2(x_i+x_{i+1}, y_i+y_{i+1})$. As a result, by tracking these movements of central points of ridges, we could assess the orders of movement of these points, *i.e.* beams' free ends.

Because the y-coordinate of each central point remained unchanged during the sliding motion of the fingertip, we focused on the x-coordinate with respect to time. Fig. 2.10(a) shows the movements of the center points of ridges during stick-to-slip transition. These findings indicate that the movements of those points are similar to those obtained during the simulation of the movements of contacting points; in which center points move gradually along the direction of slide of the fingertip. The order of sliding of those center points of Fig. 2.10(b) is similar to that during simulation Fig. 2.5(b). There are slight differences in the times of first slide, however, it is acceptable in this case. The coincidence between the simulation and experiment results thus validates our proposed model. From both results, we can conclude that, during the transition from sticking to slipping, micro localized slips on the contact line occur, propagating along the direction of slide, and the last area to move before gross slide is the area adjacent to the center zone of the contact line. This conclusion is also supported partly by previous experiments on the sliding motion of a human fingertip with epidermal ridges (see [26], or [27]). That is, the change in contact surfaces of human fingertips during sliding motion from stick to slip states (Fig. 10(a) in [26], and

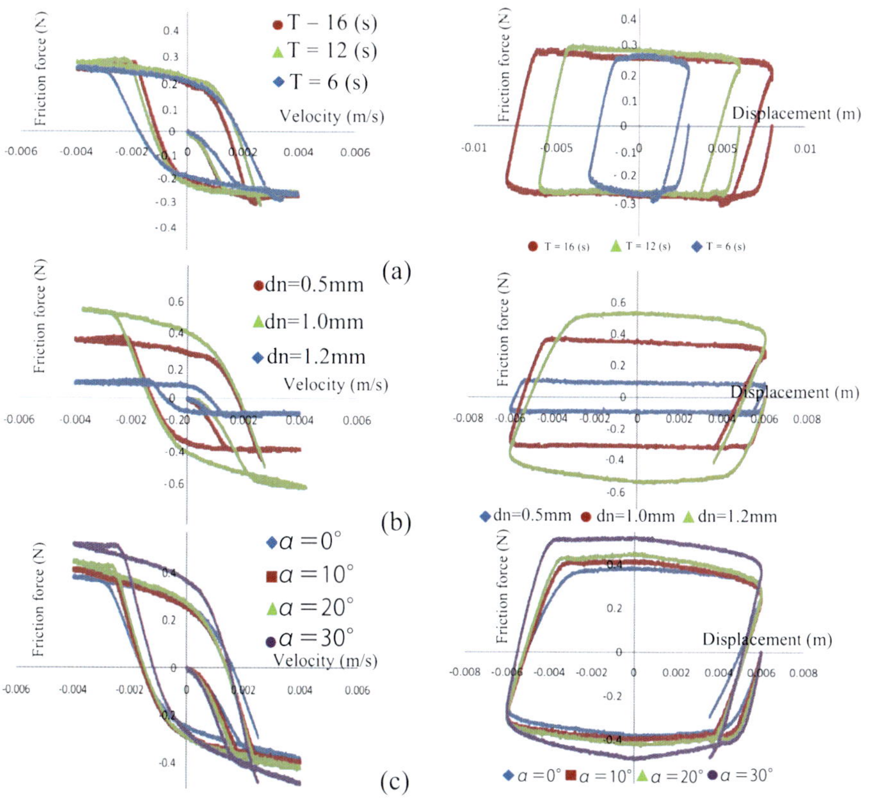

Fig. 2.12 Experimental graphes of friction-velocity and friction-displacement relations during periodical movements of the fingertip when a) Cycle time T changes. b) The contact depth d_n changes. c) The tilting angle α changes.

Fig. 6 in [27]) showed similarities to our results. Above conclusion can be used as a principle in assessing the incipient slip of a fingertip during stick-slip phase, by detecting the localized movements of contacting points. For example, if there are some proper sensors embedded inside the fingertip can detect the slip of center zone of contacting surface, it is sufficient and timely for controller to judge the slip of entire the fingertip.

In addition, we conducted experiments of the stick-slip phase when conditions vary, corresponding with content mentioned in Section 2.2.3. Fig. 2.11 showed the similarities with Fig. 2.7 during the stick-slip phase when sliding conditions change respectively.

Consequently, by conducting this experiment, one might intuitively observe the occurrence of the proposed *Localized Displacement Phenomenon* (LDP) by micro movements of contacting points on the contact surface during stick to slip transition of a sliding soft fingertip. That being said, this phenomenon dominates this transition, and by assessing this we are able to understand how the slippage happens, promising slip detection methods on soft tactile sensing system in robotics.

2.3.2 Periodical Movement Verification

As a supplement to the simulated results reported in Section 2.2.3, we assessed the responses of frictional force during periodic movements. Experiments were conducted in the same conditions with those in the simulations. Results plotted in Fig. 2.12 also reveal that the conclusions obtained from simulation results in Fig. 2.8 are verified. The results show the remarkable similarities between simulation and experiment.

2.4 Concluding Remarks

We have investigated friction characteristics of a soft fingertip in multiple phases of sliding motion. By employing model of the soft fingertip with finite number of elastic beams, and linkage spring/damper at each contact point, sliding phases of the soft fingertip have been assessed sufficiently, especially the stick-slip motion. During this phase, our model is able to extract in detail micro slips on the contact surface, including orders of their movements. In many literatures this phenomena have been found experimentally, but none of them attempted to solve this issue theoretically. By confirming this theoretically and experimentally, we can apply to real tactile soft fingertip to detect the incipient slip, in which a sensing system is used to detect slips of some important points on/near contact surface. Judgement of slip can be made if those movements are perceived. Moreover, we have also obtained characteristics of frictional force in many distinguished phases od slide. Simulation results, which matched the experimental ones, showed that this model of friction can be used describe responses of frictional force during motion of slide of the fingertip. Especially, at V_r=0, there are only some tiny sudden disturbances of friction force have been recorded, and the transitions between phases happened quite smoothly. This will allow us to use this model as a friction compensation or friction observer in position control of a sliding soft fingertip. By implementing motion of slide of the fingertip with various conditions such as changes of contact depth, cycle time, and titling angle, we acknowledge that the friction force does depend remarkably on these conditions, and consideration must be taken into account during controlling.

In this model, we have to deal with set of ODE equations to solve the localized displacements on the contact surface. The number of ODE equations depend on contact depth and titling angle. This is seemingly a limitation of our model. However, number of equation dealt in this model is quite small compared to other one. For example, the Bristle model [6] or LuGre [5] requires at least 20-25 bristles (corresponding to number of integrations) for fine results, while ours needs much less than that (9 with contact depth d_n=1 mm). Moreover, necessity of solving ODE equations is only for stick-slip transition, in which detailed movement of each contact point is requires. When the fingertip has gross slide, it is no longer necessary. This reduces remarkably complications for computation, but still assures the correctness of the model. During validation for this model we realized that when the contact depth d_n was over 2.5 mm, the stick-slip transition process was no longer matched with experimental results. It comes from the fact that, the experimental model is made from viscoelastic material which has nonlinear characteristic at high deformation. This causes the difference of normal force between our model and the real fingertip at high contact depth, resulting the dissimilarity in the stick-slip transition. This can be solved if we model each virtual beam as viscoelastic one with spring/damper element, similar to linkage element. However, we found that in reality it is not necessary to make deep contact during motion of slide. With typical application of moving an object on the surface with the fingertip, it is sufficient to implement it with light contact depth.

3

Three-Dimensional Beam Bundle Model of a Sliding Soft Fingertip

Slip, especially incipient slip, is a complicated process for soft fingertips and detection of this slip is an important factor in assuring stable manipulation for both human and robotic fingertips. By using experimental tools, previous studies have attempted to perceive this phenomenon, but none of them can explain fully the dynamic changes that occur during this process.

In the previous chapter, we proposed a 2-D dynamic model to investigate the sliding motion of semicylindrical soft fingertips on a plane with friction. However, this model cannot be exploited to describe the slip motion of soft fingertips that have complicated 3-D shapes, such as humanlike fingertips. Therefore, in this chapter, we propose a simplified yet adequate model that can act as a *platform* for the wide range modeling of fingertips. In the model, the fingertip is virtually comprised of a finite number of elastic compressible and bendable beams whose free ends act as infinitesimal contact points. The contact surface is meshed afterward using an FE method based on the coordinates of the contact points. We call this model the 3D-*Beam Bundle Model* (BBM). By introducing Coulomb's law and contact compliance into each contact point, we are able to assess the frictional characteristic during sliding motions of the fingertip. We also successfully described dynamically LDP on the contact surface during stick-slip transition, which is typical of the sliding motion of a soft fingertip. This model can be applied to different typical shapes of robotic fingertip.

3.1 Introduction

Recent robotic research has focused on the dexterous manipulation of objects using soft-fingered robotic hands, especially anthropomorphic ones. This type of research can be categorized into two main groups. The first consists of studies focusing on analyzing contact mechanics between various soft fingers and objects [10]. In the second, tactile sensing systems that imitate those of humans, along with many types of sensors, have been developed to simulate

A.-V. Ho and S. Hirai, *Mechanics of Localized Slippage in Tactile Sensing*, Springer Tracts in Advanced Robotics 99,
DOI: 10.1007/978-3-319-04123-0_3, © Springer International Publishing Switzerland 2014

human abilities in object grasping and dexterous handling [21]. Whereas the first group of studies consists primarily of analyses of stable grasping or object postures controlled by utilizing the compliance of soft fingertips during a pushing or rolling motion on the surface of objects; the second group of studies concentrate on the tactile texture perceptions of sensory fingertips in order to increase efficiency during the object manipulation processes. Of the various movements of fingertips during handling, slide/slip often occurs during any form of contact, and are considered important in dexterous manipulation [75]. For example, to assess the texture of an object's surface, a fingertip needs to slide slightly over the surface to extract information about its roughness or friction. The trend of an object to slip through the fingertips during grasping, or incipient slip, is recognized as a crucial factor in stable object manipulation. However, while the latter studies addressed all types of motion of the fingertips, such as pressing, rolling, and rubbing; the former studies have focused primarily on pure pushing and rolling movements, while ignoring slip or slide. The difficulties in modeling the sliding motion of fingertips stems from the compliance of the fingertips, their partial movements on contact surfaces, and friction force/moment. Kao and Cutkosky [12] proposed a method that combines compliance and friction on a limited surface to compute the relative sliding motion between a grasped object and soft fingers. Their paper showed concrete results in modeling contact, and in approximation of gross-motion planning. Authors in [13] introduced an algorithm for automatic stable grasping of polygonal objects with two fingers and point contact with friction. There is also research addressing pushing operations using manipulators. Yoshikawa *et al.* [29] proposed a method to identify the center of friction of an object by pushing with an unknown friction distribution. A similar method also can be found in [30], in which relevant friction parameters were estimated by performing experimental pushes and observing the resultant motion. Nevertheless, most studies only deal with quasi-static analysis, gross sliding, not on stick-slip phases (how and when they happen), or partial slip on the contact surface of soft fingertip. On the other hand, some contact models have been used in simulations, such as analytical in [31] and conventional penalty methods in [32]. While the former brings a fast and accurate description of contact, but limits penetration between objects; the latter can find the contact area, but it can only be used for penetrative contacts. An FE analysis model of the contacting of a soft fingertip was also conducted. However, one can see that this method is time consuming, and is especially solved in a static field [8]. Differing from the above approaches, authors in [33] attempted to find a best-fitted model by using the real rotational friction properties of the human fingertips; then extended the object/proxy algorithm to simulate soft finger contact.

With these considerations in mind, we have attempted to determine the frictional characteristics of a 3-D soft fingertip during unilateral sliding motion relative to an object dynamically. We have proposed a method to model vertical and horizontal deformation of a soft fingertip during sliding by introducing virtual beams. Moreover, we have modeled the contact surface by employing FE analysis methods, in conjunction with Coulomb's frictional law. This *hybrid* model, which, as previously stated, is referred to as the *Beam Bundle Model* can reduce calculation time significantly, while still assuring the dynamic behavior of the system during stick-slip transition. The results from the simulation and experimental validation explain, for the first time, theoretically *localized movements* on the contact surface during the stick-slip phase, which were mentioned experimentally in previous research.

3.2 Three-Dimensional Modeling of a Sliding Soft Fingertip

There are many shapes used in the design of robotic soft fingertips. We categorize them into two main groups (with respect to grasped objects with a rigid flat facet). The first generates a uniform distribution of normal stress, such as square or rectangular fingertips. The second features nonuniform distribution of contact force, including cylindrical, hemispherical, or humanlike fingertips. Both groups are distinguished from rigid fingertips by their compliant contact, which features a pre-slide stage and hysteresis characteristics. More particularly, with the second group, the pre-slide stage generates localized displacements on the contact surface, which plays an important role in incipient slip detection [18]. To represent those phenomena and the characteristics of a soft fingertip, a dynamic model with the appearance of friction is required. To this end, we present a model, with some simplifications, that can be employed to investigate the slip motion of soft fingertips with different shapes, especially in stick-slip transition. This model focuses on *what happens* over the entire *contact surface*, therefore we combine two methods using both discrete and FE analysis methods to model a general 3-D soft fingertip. By doing so, the computing time is reduced remarkably, while at the same time assuring the correctness of the model.

Assume that a 3-D soft fingertip with an arbitrary but continuous outer surface is pushed vertically with a specific contact penetration by a normal force F_n on a rigid flat plane, and slid horizontally with an external force F_t (Fig. 3.1(a)). [10] proposed a soft-fingertip model comprised of an infinite number of vertical elastic virtual springs that could be used to investigate the deformation of the fingertip during the pushing or rolling motion on an object. This model, however, was not suitable for describing sliding motion with the appearance of frictional force. Therefore, instead of virtual springs, we propose a model in which the soft fingertip consists of a bundle of virtual elastic beams that are compressible, tensile, and bendable (Fig. 3.1(b)). This model can represent the diverse deformations of a soft fingertip during a sliding motion, in which a soft fingertip is pushed and slid at the same time. These beams are fixed on the flat surface of the fingertip, with their free ends on the outer surface of the fingertip. Each beam has a uniform circular

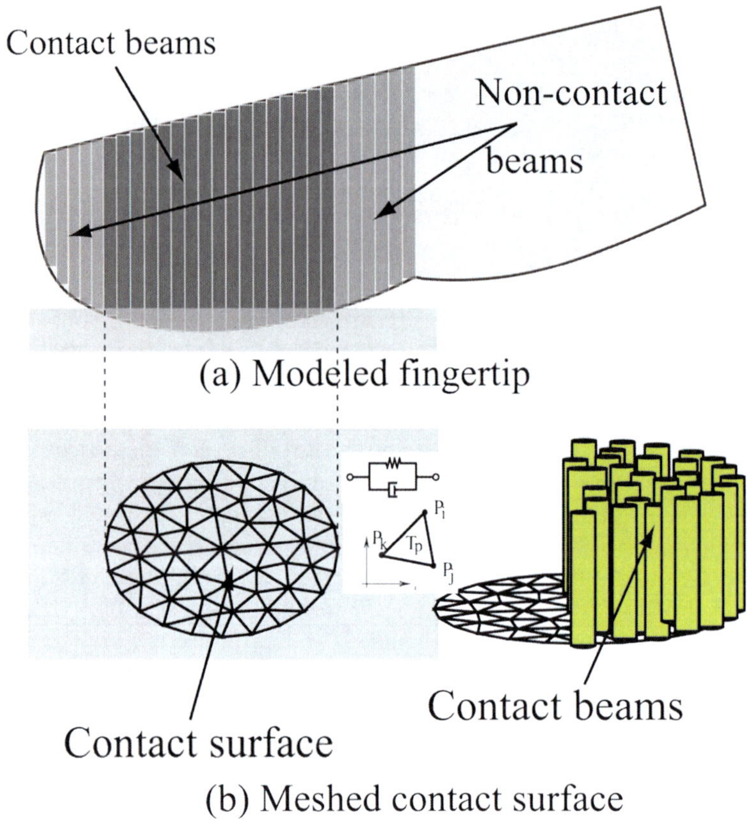

Fig. 3.1 Model of a sliding soft fingertip with virtual beams and meshed contact surface

cross-sectional area; while the lengths of the beams differ depending on their coordinates within the fingertip. When the fingertip is pushed, some beams will make contact with the object, causing them to deform. Based on the geometrical distributions, we are able to calculate the deformation of each contacting beam, forming a nonuniform distribution of normal contact on the contact surface. We then mesh the 2-D contact surface using the Voigt model to describe the elastic and viscous properties of the contact surface [25] on the $O{-}XY$ coordinates. Each node is equivalent to the free end of a contacting beam. Each element is a triangle T_p with three nodes referred to as P_i, P_j, and P_k in a counterclockwise direction. Figure 3.1(b) illustrates a possible mesh of a contact surface.

For the sake of simplicity, we made three assumptions:

1. When a beam is bent, its deformation is significant only at the free end.
2. Interactions between continuous beams only occur between their free ends on the contact surface.
3. Only beams whose free ends are acting on the contact surface are considered (dark-colored beams in Fig. 3.1(b)). Beams outside the contact surface are deemed irrelevant to the sliding motion of the fingertip (light-colored beams in Fig. 3.1(b)).

As a result, when the fingertip is pushed and slid, its deformation will be represented by deformations of all contacting beams. Moreover, external forces acting on each node on the contact surface can be assessed by calculation of the compressing and bending forces of the corresponding beam. Particularly significant, a Coulomb friction model can be introduced into each node, which was usually neglected in previous studies. Consequently, by combining both discrete and FE analysis methods into BBM of the soft fingertip, it is expected to be able to perceive the dynamic behavior of the fingertip during sliding motion, especially in the stick-slip states. The reason we used the term "Beam Bundle" is that those virtual beams are tied/constrained by the fingertip's geometry, and by an FE-based meshed contact surface in which each beam's end is attached to one contacting node. One could use other constraint methods to create interactions of a beam bundle.

3.2.1 Beam Model

For the sake of discussion, let us analyze one arbitrary beam, say the i-th beam, which has coordinates in contacting surface plane $O-XY$ as (x_i, y_i), natural length l_o^i, natural cross-sectional radius r_o^i. After being pushed at contact depth d_n^i, the normal force f_n^i acting on the free end of this beam is computed in [10] as:

$$f_n^i = k^i d_n^i = E \frac{\pi {r_o^i}^2}{l_o^i} d_n^i, \tag{3.1}$$

where E is Young's modulus. As a result, the sum of normal forces acting on all beams must be equal to F_n, as:

$$F_n = \sum f_n^i. \tag{3.2}$$

When the external tangential force F_t starts to activate, the fingertip has not been slid yet. The contact surface still sticks to the plane, causing the fingertip to deform. At this time, all contacting beams are bent at their free ends under the same bending strain δs, as illustrated in Fig. 3.2. This bending strain is calculated as originated in [34]:

$$\delta s = \frac{3\mu F_n}{16R} \frac{2-\nu}{G} \{1 - (1-\Phi)^{2/3}\}, \tag{3.3}$$

where $\Phi = F_t/\mu F_n$ is the tangential force coefficient, R is the fingertip's radius, μ is the friction coefficient, ν is Poisson's ratio, and G is the shear

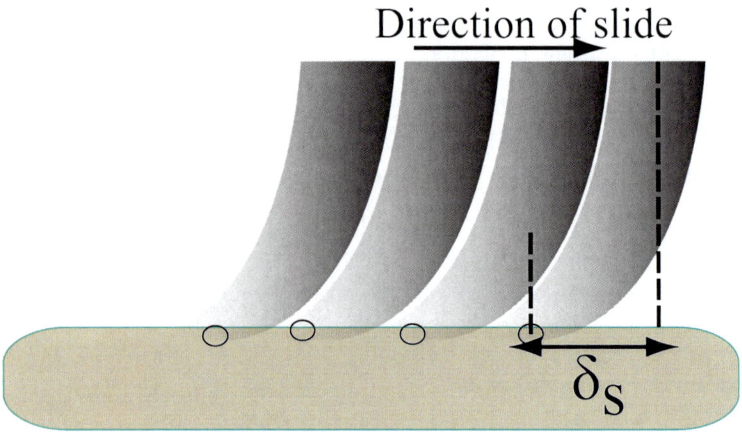

Fig. 3.2 Bending strain during stick phase

elasticity's modulus. By assessing the bending strain in Eq. 4.6, bending force f_b^i acting on the free end of the beam, with length l^i and cross-sectional area's radius r^i, is calculated in the following equation:

$$f_b^i = b^i \delta s = \frac{3EI}{(l^i)^3}\delta s = \frac{3E\pi(r^i)^4}{(l^i)^3}\delta s. \tag{3.4}$$

As a result, by employing the proposed virtual beam paradigm, we are able to estimate the two main components of *internal* force acting on the contact surface caused by deformation of the fingertip in the sliding scenario. This method allows us to compute the resulted forces quickly in a simply way given deformation of the fingertip, which is much efficient in computation time compared to a conventional FE analysis model. However, this method only represents the force acting on *one* beam, there is no cue to model natural interactions among beams using this method. Thus, the next section will introduce a solution for this issue.

3.2.2 Two-Dimensional FE Modeling of Contact Surface

There are many ways to solve FE analysis simulations of sliding contact using commercial software such as ANSYS^{TM}, or MARC^{TM}. However, these methods can neither extract the exact dynamic behavior of friction force nor micro slips during stick-slip transition. Therefore, we propose a method to calculate dynamically the output of the friction force, as well as micro sticks/slips on the contact surface during stick-slip transition.

2-D Viscoelastic Model

To derive a 2-D viscoelastic FE analysis model, we first introduce the 2-D elastic model, then extend it to a viscoelastic one.

A. 2-D Elastic Formation. Assuming that a plane region S is covered by a set of triangles as illustrated in Fig. 3.1(b). Let P_i be a nodal point with coordinates $[\xi_i, \eta_i]^T$, which belong to a triangle $\Delta P_i P_j P_k$ in the region S. Let P be an arbitrary point in the triangle $\Delta P_i P_j P_k$. Let C-$\xi\eta$ be a frame attached to the object, and (ξ, η) be the coordinates for point P. When the triangle deforms, its deformation can change the spatial coordinate of point P by $u(\xi, \eta)$ that denotes the displacement of point $P(\xi, \eta)$ along the ξ-axis, and $v(\xi, \eta)$ that denotes the displacement of point $P(\xi, \eta)$ along the η-axis. As a result, let $\mathbf{u} = [u, v]^T$ be a *displacement vector* of point P. In the region $\Delta P_i P_j P_k$, the movement of point P can be approximated by a linear combination of nodal displacements $\mathbf{u}_i = [u_i, v_i]^T$, $\mathbf{u}_j = [u_j, v_j]^T$, and $\mathbf{u}_k = [u_k, v_k]^T$ as follows:

$$
\begin{aligned}
u &= u_i N_{i,j,k} + u_j N_{j,k,i} + u_k N_{k,i,j}, \\
v &= v_i N_{i,j,k} + v_j N_{j,k,i} + v_k N_{k,i,j};
\end{aligned} \tag{3.5}
$$

where $N_{i,j,k}, N_{j,k,i}, N_{k,i,j}$ are the interpolating shape function on the triangle $\Delta P_i P_j P_k$. For example, $N_{i,j,k}$ takes 1 at point P_i and takes 0 at point P_j and P_k. This function is defined as:

$$
N_{i,j,k}(\xi, \eta) = \frac{\Delta P P_j P_k}{\Delta P_i P_j P_k} = \frac{(\xi\eta_j - \eta\xi_j) + (\xi_j\eta_k - \xi_j\eta_k) + (\xi_k\eta - \eta_k\xi)}{\Delta P_i P_j P_k}. \tag{3.6}
$$

Functions $N_{j,k,i}$, and $N_{k,i,j}$ are defined similarly by alternating indices in sequence. Let us define the deformation of the object using partial derivatives $\partial u/\partial\xi$, $\partial u/\partial\eta$, and $\partial v/\partial\xi$, $\partial v/\partial\eta$, which are calculated as follows:

$$
\frac{\partial u}{\partial \xi} = u_i \frac{\partial N_{i,j,k}}{\partial \xi} + u_j \frac{\partial N_{j,k,i}}{\partial \xi} + u_k \frac{\partial N_{k,i,j}}{\partial \xi}, \tag{3.7}
$$

$$
\frac{\partial u}{\partial \eta} = u_i \frac{\partial N_{i,j,k}}{\partial \eta} + u_j \frac{\partial N_{j,k,i}}{\partial \eta} + u_k \frac{\partial N_{k,i,j}}{\partial \eta}, \tag{3.8}
$$

$$
\frac{\partial v}{\partial \xi} = v_i \frac{\partial N_{i,j,k}}{\partial \xi} + v_j \frac{\partial N_{j,k,i}}{\partial \xi} + v_k \frac{\partial N_{k,i,j}}{\partial \xi}, \tag{3.9}
$$

$$
\frac{\partial v}{\partial \eta} = v_i \frac{\partial N_{i,j,k}}{\partial \eta} + v_j \frac{\partial N_{j,k,i}}{\partial \eta} + v_k \frac{\partial N_{k,i,j}}{\partial \eta}. \tag{3.10}
$$

The deformation can be classified into:

1. Extension along the ξ-axis denoted by $\partial u/\partial\xi$.
2. Extension along the η-axis denoted by $\partial v/\partial\eta$.
3. Shear denoted by $\partial u/\partial\eta$ and $\partial v/\partial\xi$.

We define $\epsilon_{\xi\xi}$ and $\epsilon_{\eta\eta}$ as normal strain components along ξ- and η-axes at P, and $\epsilon_{\xi\eta}$ as shear strain around point P. We then have:

$$\epsilon_{\xi\xi} = \frac{\partial u}{\partial \xi}, \qquad \epsilon_{\eta\eta} = \frac{\partial v}{\partial \eta}, \qquad 2\epsilon_{\xi\eta} = \frac{\partial u}{\partial \eta} + \frac{\partial v}{\partial \xi}. \tag{3.11}$$

Then, the strain vector, $\epsilon = \epsilon_{\xi\xi}, \epsilon_{\eta\eta}, 2\epsilon_{\xi\eta}^T$, is simply described as a linear relation with displacement as follows:

$$\epsilon = \mathbf{Lu}, \tag{3.12}$$

where L is an operator matrix. Let $\sigma_{\xi\xi}$ and $\sigma_{\eta\eta}$ are normal stress components along the ξ and η-axes at point P, and $\sigma_{\xi\eta}$ describe a shear stress component around point P. According to linear elastic material, we have the following relationship:

$$\sigma = \mathbf{D}\epsilon, \tag{3.13}$$

where $\mathbf{D}$ is referred to as an *elasticity matrix*

$$\begin{bmatrix} \lambda + 2\mu & \mu & 0 \\ \mu & \lambda + 2\mu & 0 \\ 0 & 0 & \mu \end{bmatrix},$$

where λ and μ denote Lame's constants that are described by Young's modulus E, Poisson's ratio ν, and shear elasticity modulus G as follows:

$$\lambda = \frac{\nu E}{(1+\nu)(1-2\nu)}, \qquad \mu = G = \frac{E}{2(1+\nu)}. \tag{3.14}$$

As a result, strain energy of the object over the region S is formulated as follows:

$$U = \int_S \frac{1}{2}\sigma^T \epsilon dS = \int_S \frac{1}{2}\epsilon^T \mathbf{D}\epsilon dS. \tag{3.15}$$

Exploiting the idea of the finite element method, we compute the strain energy over a set of triangles of the partitioned region S as follows:

$$U = \sum_{\Delta P_i P_j P_k} U_{i,j,k}, \tag{3.16}$$

where

$$U_{i,j,k} = \int_{\Delta P_i P_j P_k} \frac{1}{2}\epsilon^T \mathbf{D}\epsilon dS. \tag{3.17}$$

As mentioned earlier, in region $\Delta P_i P_j P_k$, the displacement u can be approximated by a linear combination of nodal displacements u_i, u_j, and u_k as follow:

$$\mathbf{u} = N_{i,j,k}\mathbf{u_i} + N_{j,k,i}\mathbf{u_j} + N_{k,i,j}\mathbf{u_k}. \tag{3.18}$$

Substituting the above equation into Eq. 3.12, strain vector ϵ is approximated as follows:

$$\epsilon = LN_{i,j,k}\mathbf{u_i} + LN_{j,k,i}\mathbf{u_j} + LN_{k,i,j}\mathbf{u_k}, \tag{3.19}$$

where

$$LN_{i,j,k} = \begin{bmatrix} \frac{\partial N_{i,j,k}}{\partial \xi} & 0 \\ 0 & \frac{\partial N_{i,j,k}}{\partial \eta} \\ \frac{\partial N_{i,j,k}}{\partial \eta} & \frac{\partial N_{i,j,k}}{\partial \xi} \end{bmatrix}. \tag{3.20}$$

Substituting Eq. 3.19 into Eq. 3.17 yields:

$$U_{i,j,k} = \frac{1}{2}\left[\mathbf{u}_i^T \ \mathbf{u}_j^T \ \mathbf{u}_i^T\right] \mathbf{K}_{i,j,k} \begin{bmatrix} \mathbf{u}_i^T \\ \mathbf{u}_j^T \\ \mathbf{u}_i^T \end{bmatrix}, \tag{3.21}$$

where $\mathbf{K}_{i,j,k}$ is referred as a stiffness matrix over $\Delta P_i P_j P_k$ that can be decomposed into two components as follows:

$$\mathbf{K}_{i,j,k} = \lambda \mathbf{J}_\lambda^{i,j,k} + \mu \mathbf{J}_\mu^{i,j,k}. \tag{3.22}$$

Matrices $\mathbf{J}_\lambda^{i,j,k}$ and $\mathbf{J}_\mu^{i,j,k}$ are typical for each triangle $\Delta P_i P_j P_k$, and calculated based on the geometrical characteristics of the triangle as follows:

$$\mathbf{J}_\lambda^{i,j,k} = \frac{h}{4\Delta P_i P_j P_k} \begin{bmatrix} \mathbf{A}_{j,k;j,k} & \mathbf{A}_{j,k;k,i} & \mathbf{A}_{j,k;i,j} \\ \mathbf{A}_{k,i;j,k} & \mathbf{A}_{k,i;k,i} & \mathbf{A}_{k,i;i,k} \\ \mathbf{A}_{i,j;j,k} & \mathbf{A}_{i,j;k,i} & \mathbf{A}_{i,j;i,j} \end{bmatrix}, \tag{3.23}$$

$$\mathbf{J}_\mu^{i,j,k} = \frac{h}{4\Delta P_i P_j P_k} \begin{bmatrix} 2\mathbf{A}_{j,k;j,k} + \mathbf{C}_{j,k;j,k} & 2\mathbf{B}_{j,k;k,i} + \mathbf{C}_{j,k;k,i} & 2\mathbf{B}_{j,k;i,j} + \mathbf{C}_{j,k;i,j} \\ 2\mathbf{A}_{k,i;j,k} + \mathbf{C}_{k,i;j,k} & 2\mathbf{B}_{k,i;k,i} + \mathbf{C}_{k,i;k,i} & 2\mathbf{B}_{k,i;i,j} + \mathbf{C}_{k,i;i,j} \\ 2\mathbf{B}_{i,j;j,k} + \mathbf{C}_{i,j;j,k} & \mathbf{B}_{i,j;k,i} + \mathbf{c}_{i,j;j,i} & 2\mathbf{B}_{i,j;i,j} + \mathbf{C}_{i,j;i,j} \end{bmatrix}, \tag{3.24}$$

where

$$\mathbf{A}_{i,j;l,m} = \begin{bmatrix} (\eta_i - \eta_j)(\eta_l - \eta_m) & -(\eta_i - \eta_j)(\xi_l - \xi_m) \\ -(\xi_i - \xi_j)(\eta_l - \eta_m) & (\xi_i - \xi_j)(\xi_l - \xi_m) \end{bmatrix}$$

$$\mathbf{B}_{i,j;l,m} = \begin{bmatrix} (\eta_i - \eta_j)(\eta_l - \eta_m) & 0) \\ 0 & (\xi_i - \xi_j)(\xi_l - \xi_m) \end{bmatrix}$$

$$\mathbf{C}_{i,j;l,m} = \begin{bmatrix} (\xi_i - \xi_j)(\xi_l - \xi_m) & -(\xi_i - \xi_j)(\eta_l - \eta_m) \\ -(\eta_i - \eta_j)(\xi_l - \xi_m) & (\eta_i - \eta_j)(\eta_l - \eta_m) \end{bmatrix}$$

Synthesizing over region S, the stiffness matrix can be rewritten as follows:

$$\mathbf{K} = \lambda \mathbf{J}_\lambda + \mu \mathbf{J}_\mu, \tag{3.25}$$

where $\mathbf{J}_\lambda$ and $\mathbf{J}_\mu$ are referred to as connection matrices. These connection matrices depend solely on the geometric quantities, or the coordinates of nodal points, and are calculated by incorporating matrices $\mathbf{J}_\lambda^{i,j,k}$ and $\mathbf{J}_\mu^{i,j,k}$ of each triangle based on the contribution of each triangle to the whole mesh of region S.

As a result, strain energy over region S can be rewritten as follows:

$$U = \frac{1}{2}\mathbf{u}^T\mathbf{K}\mathbf{u}, \tag{3.26}$$

where u denotes the nodal displacement vector of partitioned region S. Taking the derivative of the above strain energy relative to vector u, we have the formulation of a set of elastic forces generated on all nodal points as:

$$\mathbf{F}_{ela} = \mathbf{K}\mathbf{u} = (\lambda\mathbf{J}_\lambda + \mu\mathbf{J}_\mu)\mathbf{u}. \tag{3.27}$$

Consequently, we have reported an attempt to model a 2-D elastic region exploiting FE analysis. Next, we extend the elastic formulation to dynamic viscoelastic formulation for use in modeling the contact surface of a sliding soft fingertip.

B. 2-D Viscoelastic Formation. As mentioned, the stress-strain relationship of an elastic object can be specified by a constant E that is Young's modulus. The 2-D isotropic deformation in Eq. 3.13 is rewritten as follows:

$$\sigma = (\lambda\mathbf{I}_\lambda + \mu\mathbf{I}_\mu)\epsilon, \tag{3.28}$$

where

$$\mathbf{I}_\lambda = \begin{bmatrix} 1 & 1 & 0 \\ 1 & 1 & 0 \\ 0 & 0 & 0 \end{bmatrix}, \qquad \mathbf{I}_\mu = \begin{bmatrix} 2 & 0 & 0 \\ 0 & 2 & 0 \\ 0 & 0 & 1 \end{bmatrix}. \tag{3.29}$$

To derive formations for viscoelastic material, let us recall the stress-strain relationship of a viscoelastic object by an operator:

$$E + c\frac{d}{dt}. \tag{3.30}$$

From the above observation, replacing Eq. 3.28 by two viscoelastic operators will yield 2-D isotropic viscoelastic deformation as follows:

$$\lambda = \lambda^{ela} + \lambda^{vis}\frac{d}{dt}, \qquad \mu = \mu^{ela} + \mu^{vis}\frac{d}{dt}. \tag{3.31}$$

Two constants λ^{ela} and μ^{ela} specify the elasticity of the object while λ^{vis} and μ^{vis} describe its viscosity. The operator $\lambda^{ela} + \lambda^{vis}\frac{d}{dt}$ characterizes normal viscoelasticity of the object while its shear viscoelasticity is represented in $\mu = \mu^{ela} + \mu^{vis}d/dt$. Having this relationship is helpful in the formation of the viscoelastic force acting on region S. Putting Eq. 3.31 into Eq. 3.27 yields:

$$\mathbf{F}_{viscoelastic} = (\lambda^{ela} + \lambda^{vis}\frac{d}{dt}\mathbf{J}_{\lambda} + \mu^{ela} + \mu^{vis}\frac{d}{dt}\mathbf{J}_{\mu})\mathbf{u}. \tag{3.32}$$

Replacing $d\mathbf{u}/dt$ by $\mathbf{v} = \dot{\mathbf{u}}$ and by rearranging the above equation, we finally obtain the formation of the viscoelastic force as follows:

$$\mathbf{F}_{ve} = (\lambda^{ela}\mathbf{J}_{\lambda} + \mu^{ela}\mathbf{J}_{\mu})\mathbf{u}_N + (\lambda^{vis}\mathbf{J}_{\lambda} + \mu^{vis}\mathbf{J}_{\mu})\mathbf{v}_N, \tag{3.33}$$

or

$$\mathbf{F}_{ve} = \mathbf{K}_{ela}\mathbf{u}_N + \mathbf{K}_{vis}\dot{\mathbf{u}}_N. \tag{3.34}$$

C. Dynamic 2-D Deformation Formation. To formulate dynamic deformation for a 2-D object, we formulate the kinetic energy of the object to derive the inertial forces applied to it. Let ρ be the density of an object at point $\mathrm{P}(\xi, \eta)$. Similar to the derivation of strain energy, the total kinetic energy on the region S is simply given by:

$$T = \int_S \frac{1}{2}\rho\dot{\mathbf{u}}^T\dot{\mathbf{u}}ds, \tag{3.35}$$

or

$$T = \sum_{\Delta P_i P_j P} T_{i,j,k}, \tag{3.36}$$

where

$$T = \int_{\Delta P_i P_j P_k} \frac{1}{2}\rho\dot{\mathbf{u}}^T\dot{\mathbf{u}}ds \tag{3.37}$$

provides partial kinetic energy in region $\Delta P_i P_j P_k$. Following a similar method for the derivation of strain energy, the total kinetic energy can be described by a quadratic form of the nodal velocity vector as follows:

$$T = \frac{1}{2}\dot{\mathbf{u}}_N^T\mathbf{M}\dot{\mathbf{u}}_N, \tag{3.38}$$

where $\mathbf{M}$ is referred to as an *inertia matrix* that is synthesized from partial matrices over triangles of the partitioned region S, such as:

$$M_{i,j,k} = \rho\frac{\Delta P_i P_j P_k}{12}\begin{bmatrix} 2\mathbf{I}_{2\times 2} & \mathbf{I}_{2\times 2} & \mathbf{I}_{2\times 2} \\ \mathbf{I}_{2\times 2} & 2\mathbf{I}_{2\times 2} & \mathbf{I}_{2\times 2} \\ \mathbf{I}_{2\times 2} & \mathbf{I}_{2\times 2} & 2\mathbf{I}_{2\times 2} \end{bmatrix}. \tag{3.39}$$

Details of above derivations can be seen in the Appendix A. To form dynamic motion equation of nodal points on the region S, we introduce a Lagrangian as follows:

$$L = T - U + W, \tag{3.40}$$

where W is the work of an applied external force. From this equation, dynamic motion equation can be derived.

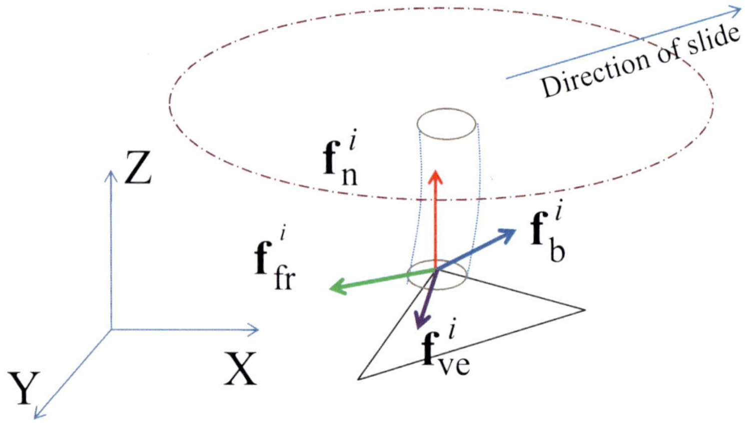

Fig. 3.3 Analysis of force acting on an arbitrary beam

BBM Model of Soft Fingertip

As a subsequent attempt at the formation of motion equation for a sliding fingertip we implement a combination of beam paradigm and the FE analysis, finally extracting the micro movements of nodal points on the contact surface of the soft fingertip. We first incorporate the acting forces that are analyzed below into one contacting node. These forces are then put into Eq. 3.40 as external forces. As a result, the motion equations can be perceived.

Assuming that one beam is bent due to the movement of the fingertip. This beam's free end is attached to one node, say i of one triangle, say $\Delta P_i P_j P_k$, on the partitioned contact surface. Analysis of applied force is illustrated in Fig. 3.3. For each node, in addition to viscoelastic force $\mathbf{f}_{ve}^{i}$ calculated implicitly by Eq. 3.34, there exist friction force $\mathbf{f}_{fr}^{i}$, normal force $\mathbf{f}_{n}^{i}$, and bending force $\mathbf{f}_{b}^{i}$ (Fig. 3.3).

The friction forces conform to Coulomb's law as they oppose the motion of contact, and if there is no motion at the contact, the friction force can act in any direction with any magnitude less than or equal to the product of the friction coefficient and the normal force. To be capable to assess stick/slip motion of each contact node, slip condition is proposed based on the Coulomb condition and the constraint stabilization method (CSM). This method provides a numerical computation for a system of differential equations under geometric constraints [see Appendix B.2]. When applying for a node, CSM assures the fixture of that node, if it keeps sticking to the contact facet. Let $\mathbf{A}_{2,2}^{i}$ be a matrix to describe the constraint of the i-th node. When the node is fixed, the constraint matrix is specified as:

$$\mathbf{A}_{1}^{i} = \begin{bmatrix} 1 & 0 \\ 0 & 1 \end{bmatrix}. \tag{3.41}$$

As a result, the boundary condition for the fixed contacting point is:

$$\mathbf{A}^i\mathbf{u}^i = \mathbf{0}, \tag{3.42}$$

where $\mathbf{u}^i$ is referred as displacement vector of the i-th node. When the node gives way in any direction, *i.e.* unconstrained, then $\mathbf{A}^i = \mathbf{0}_{2,2} = \mathbf{A}_0^i$. Taking all above consideration, the stick/slip conditions of a contacting node are described as follows:

$$\mathbf{f}^i_{fr} = \begin{cases} \mathbf{f}^i_{ve} + \mathbf{f}^i_b, & \mathbf{f}^i_{ve} + \mathbf{f}^i_b < \mu\mathbf{f}^i_n \Rightarrow Stick \Rightarrow \mathbf{A}^i = \mathbf{A}^i_1 \\ \mu\mathbf{f}^i_n, & \mathbf{f}^i_{ve} + \mathbf{f}^i_b \geq \mu\mathbf{f}^i_n \Rightarrow Slip \Rightarrow \mathbf{A}^i = \mathbf{A}^i_0 \end{cases} \tag{3.43}$$

For each iteration, the stick state or the slip state of each contacting node is checked based on the value of friction force, to determine the value of the constrain matrix $\mathbf{A}^i$. Finally, constraint $2N \times 2N$ matrix $\mathbf{A}$ of all nodes of the contact surface will be synthesized from partial constraint matrices as

$$\mathbf{A}^T = \begin{bmatrix} \mathbf{A}^0 & \mathbf{0} & \dots & \mathbf{0} \\ \mathbf{0} & \mathbf{A}^1 & \dots & \mathbf{0} \\ \vdots & \vdots & \ddots & \vdots \\ \mathbf{0} & \mathbf{0} & \dots & \mathbf{A}^N \end{bmatrix}, \tag{3.44}$$

with N being the number of contacting nodes. As a result, the geometric constraint is described in the following equation:

$$\Phi^H = \mathbf{R}\triangle\mathbf{A}^T\mathbf{u}_N = \mathbf{0}. \tag{3.45}$$

We apply CSM to incorporate this constraint into dynamic equations. Let us define a critical damping of the constraint:

$$\ddot{\mathbf{R}} + 2\omega\dot{\mathbf{R}} + \omega^2\mathbf{R} = \mathbf{0}, \tag{3.46}$$

where ω denotes a predetermined angular frequency, which takes a large positive value. Then, the quantity R converges to zero quickly, *i.e.* the boundary constraint in Eq. 3.45 is satisfied during the computation. Computing the above constraint stabilization equation yields:

$$\mathbf{A}^T\ddot{\mathbf{u}}_N + \mathbf{A}^T(2\omega\dot{\mathbf{u}}_N + \omega^2\mathbf{u}_N) = \mathbf{0}. \tag{3.47}$$

Consequently, Lagrangian in Eq. 3.40 under a set of geometric constraints $\mathbf{R}$ is reformulated as:

$$\mathbf{L} = \mathbf{T} - \mathbf{U} + \mathbf{W} + \lambda^T\mathbf{A}^T\mathbf{u}. \tag{3.48}$$

Therefore, a set of motion equations of all nodal points is formulated as:

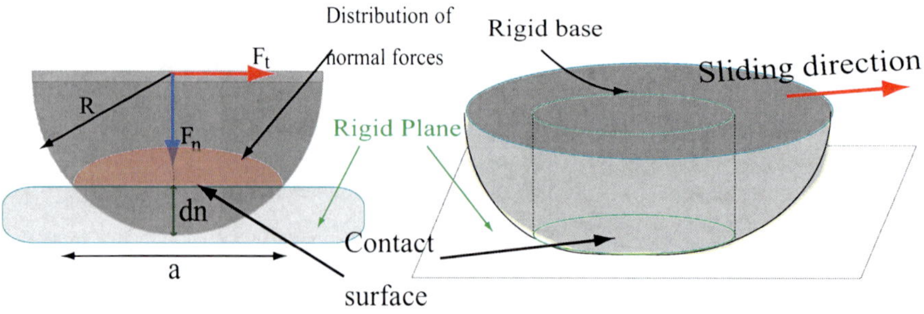

Fig. 3.4 Model of a hemispherical soft fingertip

$$-\mathbf{M}\ddot{\mathbf{u}}_N - \mathbf{K}_{ela}\mathbf{u}_N - \mathbf{K}_{vis}\dot{\mathbf{u}}_N + \mathbf{F} + \mathbf{A}\lambda = \mathbf{0}, \tag{3.49}$$

where $\mathbf{M}$ is the inertia matrix of the 2-D FE analysis contact surface, $\mathbf{F} = \mathbf{F}_{fr} + \mathbf{F}_b$ being the external force vector on the contact surface, and λ being a set of Lagrange multipliers. Recalling Eq. 3.47 and introducing the relation $\mathbf{v}_N = \dot{\mathbf{u}}_N$, equation motions of all nodes can be described as follows:

$$\begin{cases} \mathbf{v}_N = \dot{\mathbf{u}}_N \\ \mathbf{M}\dot{\mathbf{v}}_N - \mathbf{A}^T\lambda = -\mathbf{K}_{ela}\mathbf{u}_N - \mathbf{K}_{vis}\mathbf{v}_N + \mathbf{F} \\ -\mathbf{A}^T\dot{\mathbf{v}_N} = \mathbf{A}^T(2\omega\mathbf{v}_N + \omega^2\mathbf{u}_N) \end{cases} \tag{3.50}$$

Namely

$$\begin{bmatrix} \mathbf{I} & \mathbf{0} & \mathbf{0} \\ \mathbf{0} & \mathbf{M} & -\mathbf{A} \\ \mathbf{0} & -\mathbf{A}^T & \mathbf{0} \end{bmatrix} \begin{bmatrix} \dot{\mathbf{u}}_N \\ \dot{\mathbf{v}}_N \\ \lambda \end{bmatrix} = \begin{bmatrix} \mathbf{v}_N \\ -\mathbf{K}_{ela}\mathbf{u}_N - \mathbf{K}_{vis}\mathbf{v}_N + \mathbf{F} \\ \mathbf{A}^T(2\omega\mathbf{v}_N + \omega^2\mathbf{u}_N) \end{bmatrix}. \tag{3.51}$$

This equation is linear and solvable since the matrix is regular, implying that we can compute $\dot{\mathbf{u}}_N$, and $\dot{\mathbf{v}}_N$. Details of numerical integration of ordinary differential equations can be seen in Appendix B.

Consequently, by using a combination of equations for virtual beams and equations of motion for the contact surface, we can perceive the dynamic friction force acting on each node during its stick/slip, as well as the total friction force on the contact surface. Moreover, a dynamic look at the contact surface during stick-to-slip phase is also capable to assess.

Table 3.1 Parameter Values Used in Simulation of Hemispherical Fingertip

Parameter	Value	Unit
R	10	mm
d_n	2	mm
E	0.623	MPa
ν	0.48	
C	100	Pa.s
μ	0.9	

3.3 Simulations

To conduct the simulation, we implemented the model in a C++ programming environment to optimize the calculation time. To solve Eq. 3.51, we employed a sixth-order Runge-Kutta method with a sampling time of 1.0x10^{-6} s (see Appendix B). The simulation runs on a standard PC with a 1.8 GHz processor.

In this simulation, the soft fingertip has a diameter of 10.0 mm, which is similar to an adult thumb's fingertip; and it is pushed at a contact depth d_n and moved with constant velocity v. The contact surface has a circular form, with diameter a being calculated as follows:

$$a = \sqrt{R^2 - (R - d_n)^2}. \tag{3.52}$$

Because the fingertip can slide in any direction, the mesh of the contact surface must be symmetrical to the centroid of the contact surface. One example of meshed contact surface can be seen in Fig. 3.1. The number of contacting nodes is N, as is the number of contact beams. The normal force acting on a node on the contact surface can be estimated using the modified form of Eq. 3.1:

$$f_n^i = k^i d_n^i = E\frac{\pi {r_o^i}^2}{l_o^i}\{\sqrt{R^2 - (x_i^2 + y_i^2)} - (r - d_n)\}. \tag{3.53}$$

Other parameters are summarized in Table 3.1. This simulation focuses on stick-to-slip transition, which is the transient phase prior to the stable slide of the fingertip. By solving Eq. 4.11, the transient period of stick-slip of the moving hemispherical fingertip can be perceived. Some results are reported below.

3.3.1 Friction Response

Figure 3.5 shows a plot of friction force acting on a contact surface during the stick-to-slip phase. It is calculated by totaling all the friction forces acting

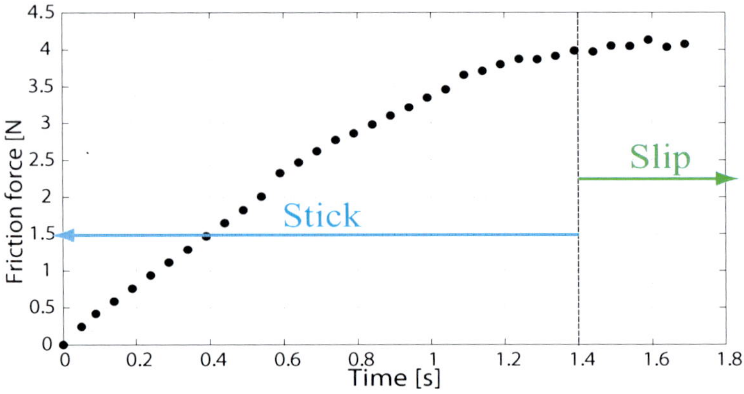

Fig. 3.5 Friction force during one stick-to-slip transition trial

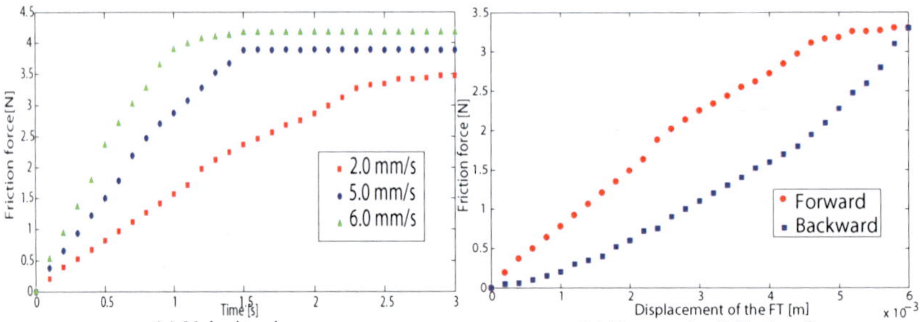

(a) Velocity changes (b) Hysteresis of friction force

Fig. 3.6 Responses of friction force of simulation trials

on the nodes of the contact surface. It is easy to realize two different stages of friction force during this phase. In the stick stage, the friction force keeps increasing; while it is unchanged in the slide phase. There is no sudden change between the two stages; it occurs smoothly. This phenomenon occurs only with ductile material, like that of a soft fingertip; but it is unrealistic for solid objects when the body accelerates, even although the external forces on the body are less than the threshold of frictional force [6]. When the sliding velocity varies, the friction responses also change correspondingly, as plotted in Fig. 3.6(a); in a way that the faster the velocity is, the faster the moment of slide occurs. One of the typical characteristics of friction of soft material is the hysteresis of friction prior to a slide is also perceived in this simulation. The fingertip is moved until the moment immediately before the slide regime, stopped, and moved back to the original state. The response of friction over displacement of the fingertip is recorded and plotted in Fig. 3.6(b), showing that the forward and backward friction force during the stick phase make a

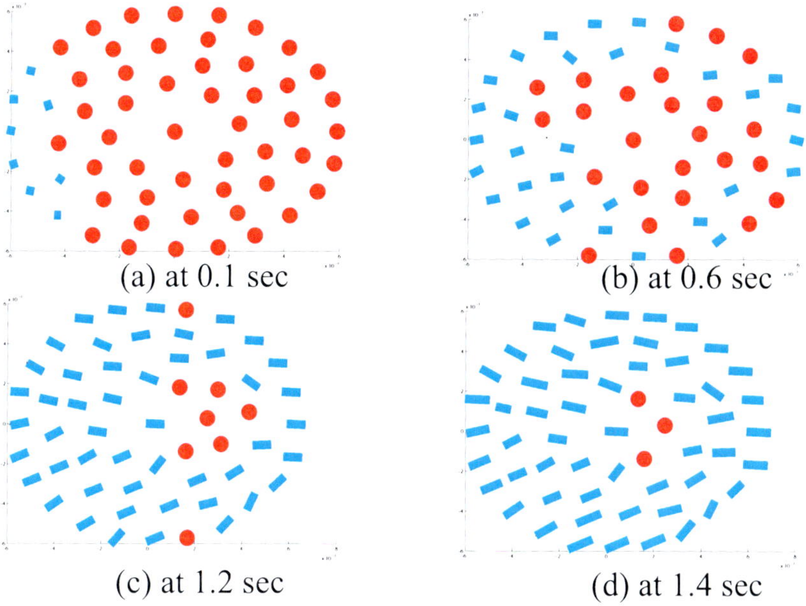

Fig. 3.7 The simulation's micro displacements on the contact surface during stick-to-slip transition. Red dots indicate stick contact nodes, while blue bars show trajectories of slipped contact nodes.

hysteresis. The gap of the hysteresis depends on sliding velocity as well. This behavior agrees with the experimental results of [5], which has shown that friction behaves like a spring if the applied force is less than the break-away force.

3.3.2 Localized Slippage during Stick-to-Slip

In the previous chapter, we identify the micro movements of contacting points during the transition from the stick state to the slip state of a 2-D soft fingertip. We also attempt to observe this phenomenon in a 3-D model simulation. By illustrating the partial trajectories of contact nodes during the stick-to-slip transition in Fig. 3.7, we can get a close look at the micro changes in strain on the contact surface. That is, the peripheral area of the contact surface slips first (Fig. 3.7(a)); the slipped area propagates to the center of the contact surface in Fig. 3.7(b) to (c). This is well known as the incipient slip of a hemispherical soft fingertip, and has been mentioned in many studies [21] and [8]. In addition, [18] also found out experimentally that during the stick-to-slip transition the center of the contact surface was the last one to give way right before the gross slippage of the moved fingertip on the object's plane.

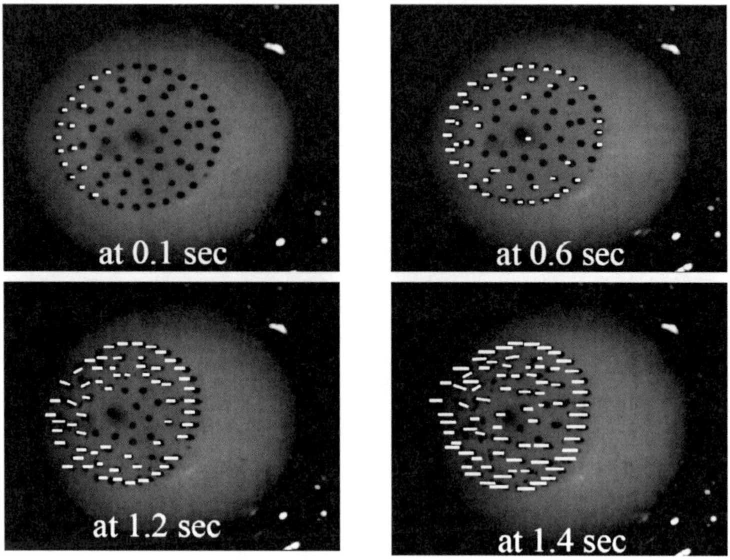

Fig. 3.8 Micro displacements on the contact surface during stick-to-slip transition. White bars show the trajectories of slipped contact nodes.

As a result, the transition from stick phase to totally slipped phase of a sliding soft fingertip happens in a way that might be surprising. There actually *exist* micro movements on the contact surface even although the fingertip itself sticks relatively to the plane. This must change our thinking about slip perception of a soft fingertip. Instead of making an observation of frictional change as the conventional method did, which is not timely enough in assessing slippage, one could propose methods to predict the micro movements on the contact surface, *i.e.* LDP, that promises to an efficient way to detect when the fingertip slips. In conjunction with emerging technologies in sensor fabrication, the potential for the applications of LDP is promising.

As stated above, simulation trials were implemented in a C++ environment, and on a standard PC. We are able to conduct a simulation of a sliding cylindrical fingertip model in real time. For a model of a hemispherical fingertip, each trial took 4 to 10 minutes to finish, depending on the initial conditions such as sliding velocity or contact depth. We are attempting to speed up the simulation trials by utilizing the power of parallel computation, using CUDA in GNU-equipped computers. It is very promising that we will be able to implement simulations in real time in the near future. In comparison with solid FE analysis, a simulation implemented in commercial software was notably quicker. The solid FE analysis simulation, apparently, brings detailed stress-strain information at all nodes, but our model can describe stick-slip transition dynamically with promising applications for real-time use.

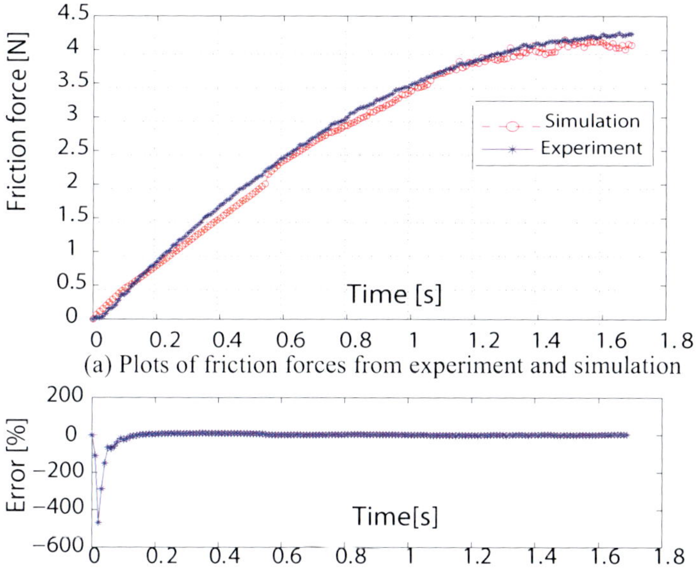

(a) Plots of friction forces from experiment and simulation
(b) Error between two plots

Fig. 3.9 Comparison between simulation and experimental friction force at $v = 5.0\,\mathrm{mm/s}$

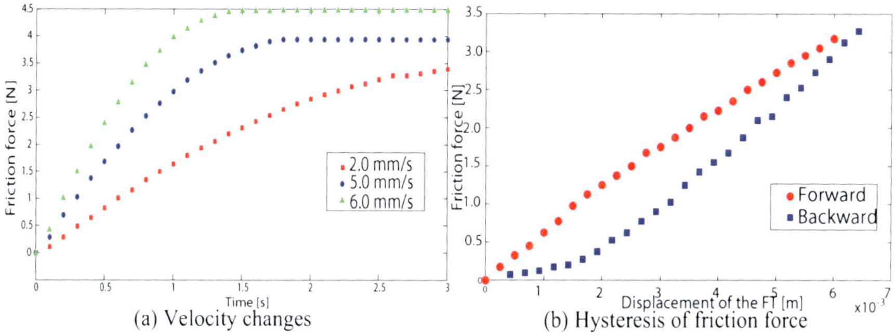

(a) Velocity changes
(b) Hysteresis of friction force

Fig. 3.10 Experimental friction force during stick-to-slip transition

3.4 Experimental Validation

To verify the simulated model, we conducted an experiment in which one polyurethane rubber soft fingertip was moved over a rigid surface by a 2-DOF stage. A 3-DOF load cell was attached to the fingertip to measure the external forces acting on it. On the outer surface of the fingertip, there are marking dots used to emulate contacting nodes. These dots have similar coordinates to those of the contacting nodes in the simulation. To create these dots, 0.5 mm radius holes distributed similarly to those of the nodes on the meshed contact surface were drilled in a thin steel sheet. This sheet was pushed downward on top of the soft fingertip with the same contact depth d_n as in the simulation. We used spray paint to create the black dots through the drilled holes, as illustrated in Fig. 3.8. A high-speed camera was used to track the movements of these dots. The white bars in Fig. 3.8 illustrate localized displacements of tracked dots on the contact surface during the stick phase over time. We also determined that localized slips on the contact surface occur first at the outward border, and then propagate into the center zone. As a result, the localized displacement phenomenon is verified experimentally, showing a similar tendency to the simulation result shown in Fig. 3.7. Thus, we can insist upon the existence of LDP that dominates the transition from stick state to slip state of a soft fingertip, showing that by the observation of micro movements of contacting points on the contact plane one can assess *how*, and especially *when* the total slippage happens. This will help robotic control systems to realize/judge the occurrence of incipient slip between fingers and a grasped object, which is considered crucial in stable manipulation.

Figure 3.9(a) compares simulated and experimental friction forces during stick-to-slip transition with $v = 5.0\,\text{mm/s}$. We observe very good agreement between the two results. The relative error is large at the very beginning moment of the slide, but quickly vanishes, as illustrated in Fig. 3.9(b). The change in friction force when the velocity of slide varies is illustrated in Fig. 3.10(a). When the velocity increases, overt slip occurs more quickly. In particular, we can determine by looking at saturated values that when the velocity increases, the friction coefficient increases correspondingly. This is due to the viscosity characteristic of the friction force, in which the friction coefficient depends on the sliding velocity [25]. This also is represented in the simulation results, as mentioned in Section 3.3.1. Hysteresis of friction force and fingertip displacement is plotted in Fig. 3.10(b), showing the agreement with simulation's result in Fig. 3.6(b).

3.5 Discussion

Both our simulation and experimental results confirmed the existence of LDP on the contact area during transition from stick to slip. This phenomenon dominates the transition of soft objects before sliding, and is important in understanding slip perception. The existence of this phenomenon suggests that a perception of micro displacements on the contact surface by sensing systems, either directly or in a synthesized way, may result in a more timely assessment of slippage of soft objects. Thus, slippage may be detected at its onset, or even *before*. This is important for slip prevention in robotic applications such as hand manipulation and human-robot interactions. In the next section, we present some applications of LDP in assessing the slip perception of several soft tactile systems.

3.5.1 Stick-Slip Transition

This section presents our proposal for a *BBM* platform using an FE analysis method, which enables a simulation of the dynamic behavior of a sliding soft fingertip. We believe that this model can be used to simulate other types of soft fingertips, even those with complex surfaces. Unless it is impossible to perceive the distribution of normal forces acting on the contact surface, stick-slip transitions and localized displacements can be assessed using our model. This model can also be used to explain the sliding of human fingertips, supported in part by the results of previous experiments on the sliding motion of a human fingertip with epidermal ridges [26] and [27]. In those studies, if we take a look at how the contact surfaces of human fingertips change during their sliding motion from stick to slip states in [26], [27]. In particular, in Fig. 4 of [27], there are similarities to our simulation results; in which localized displacements occur first at the border zones, propagating to the inward zone. As a result, our model promises a platform that is capable of dynamically representing the sliding motion of many soft fingertips, ranging from robotic to human ones. We have used magnetic resonant images and a snake algorithm to specify the exact shape of fingertips, as well as the locations of bone and other tissues. Based on this information, the stiffness of each beam, corresponding to bone, soft tissue, etc., can be determined. Physical connections between beams of different stiffness are made to assure the flawlessness of the model. Contact surfaces are meshed, corresponding to the characteristics of human skin. Thus, by extending our proposed model, we can assess the simulation of the sliding of a human finger.

3.5.2 Friction Torsion

We investigated bilateral movements of soft fingertips, yet complex one. It was necessary to introduce friction torque into the model, as stated in [2]. To analysis this movement, we used the foundation of an instantaneous center of rotation (i.c.r., which has been mentioned in [12]. According to the illustration in Fig. 3.11, moment about the z-axis, or normal to the contact area, is calculated as:

$$m = -\sum_{i}^{N} m^i = -\sum_{i}^{N} \mathbf{r}^i \times \frac{\mathbf{v}^i}{\| \mathbf{v}^i \|} f_{fr}^i, \tag{3.54}$$

with f_{fr}^i be the friction force at the contacting node i calculated in Eq. 3.43; $\mathbf{v}^i$ is the velocity vector with respect o the i.c.r.; and $\mathbf{r}^i$ is the location vector of the i-th node.

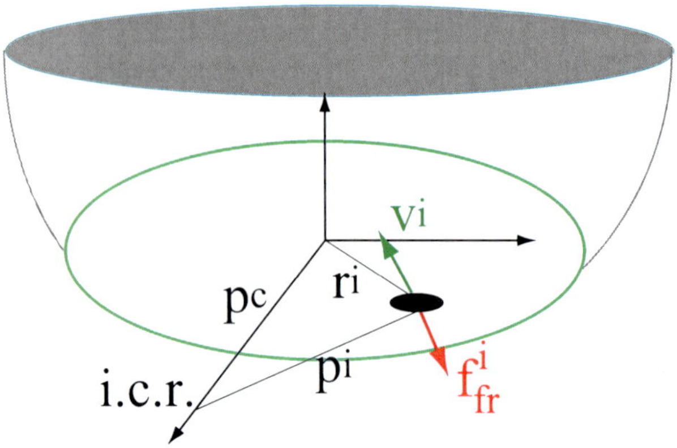

Fig. 3.11 Contact surface and i.c.r.

It is obvious that by using this hypothesis more constraints need to be introduced for all contacting nodes of this model. With respect to i.c.r., at each iteration the direction of movement of each contact node is considered perpendicular to $\mathbf{p}^i$, *i.e.*

$$Q^i = \mathbf{v}^i \cdot \mathbf{p}^i = v^i_x p^i_x + v^i_y p^i_y = 0. \tag{3.55}$$

This is a nonholonomic Pfaffian constraint. Let $\mathbf{Q}$ be the matrix description of Eq. 3.55 for all contact nodes. Since $\mathbf{Q}$ contains velocity components, any differential equation with second-order derivatives of $\mathbf{Q}$ is not available. Thus, we apply the following differential equation:

$$\dot{\mathbf{Q}} + \nu \mathbf{Q} = 0, \tag{3.56}$$

where ν is a predetermined damping ratio. In companion with the holonomic constraint described in Eq. 3.55, we can incorporate these constraints into equations of motion as [44]

$$\begin{bmatrix} \mathbf{I} & \mathbf{0} & \mathbf{0} & \mathbf{0} \\ \mathbf{0} & \mathbf{M} & -\mathbf{\Phi}^{HT} & -\mathbf{\Phi}^{NT} \\ \mathbf{0} & -\mathbf{\Phi}^{H} & \mathbf{0} & \mathbf{0} \\ \mathbf{0} & -\mathbf{\Phi}^{N} & \mathbf{0} & \mathbf{0} \end{bmatrix} \begin{bmatrix} \dot{\mathbf{u}}_N \\ \dot{\mathbf{v}}_N \\ \lambda^H \\ \lambda^N \end{bmatrix} = \begin{bmatrix} \mathbf{v}_N \\ -\mathbf{K}_{ela}\mathbf{u}_N - \mathbf{K}_{vis}\mathbf{v}_N + \mathbf{F} \\ \gamma^H \\ \gamma^N \end{bmatrix}. \tag{3.57}$$

In this equation, $\mathbf{\Phi}^H$ and $\mathbf{\Phi}^N$ are holonomic and nonholonomic constraint matrices calculated from constraints $\mathbf{R}$ (Eq. 3.45) and $\mathbf{Q}$ (Eq. 3.55) respectively according to [44]; λ^H and λ^N are sets of Lagrange multipliers; γ^H and

γ^N are derived from Eq. 3.46 and Eq. 3.56. By numerically integrating these equations, responses of friction force and friction torque can be assessed.

3.6 Concluding Remarks

This chapter presented a BBM with a FE analysis method of discretely represented virtual beams, which simulate the dynamic behavior of a sliding soft fingertip. We focused on the stick-slip transition of sliding, which is considered an important factor in stable grasping. We successfully reproduced in our simulation dynamic localized displacements of the contact surface during stick-slip transition, which also was determined experimentally. By performing validation experiments, we concluded that our model was sufficient to dynamically explain this complex phase of a sliding soft fingertip.

4

Modeling of a Sliding Human Fingertip

In this thesis, we employed our proposed Beam Bundle Model (BBM) (see previous chapters) in the attempted modeling of a human fingertip. In order to implement this, we utilized magnetic resonance imaging (MRI) of a subject's fingertip to construct a mathematical model of the structure of a fingertip. We first characterize pre-sliding regime on a real human finger in order to find out which factors is the most important in term of haptic display. We succeeded in creating a representation of localized displacements (also referred to as local skin stretch) during the pre-slide phase of the fingertip, which is considered crucial to assessing stick/slip events on the contact surface during contact with the outside world. The results of this research can be utilized in further studies in haptic sensation, and to develop sensors for the detection of slippage.

4.1 Introduction

Of the many senses, tactile perception is considered to one of the most important for both humans and robots. It permits the acquisition of information from the outside world, especially with respect to the characteristics of an object being grasped so as to enable stable/dexterous manipulation. Research into tactile perception in robotics has been conducted for more than 30 years, mainly focusing on mimicking aspects of the human fingertip ranging from shape, function and structure through to the distribution of mechanoreceptors [45]. With respect to this, it is vital for robotics researchers to understand the functions and characteristics of the anatomy of the human fingertip. Among the multiple modalities that the human finger can sense, temperature, force, vibration, *etc.*, deformation of the skin plays an important role in touch sensation. It stimulates four types of mechanoreceptor cells lying just beneath the skin, and these stimulated cells then send signals to the brain for processing. Recently, there have been many studies that have

A.-V. Ho and S. Hirai, *Mechanics of Localized Slippage in Tactile Sensing*,
Springer Tracts in Advanced Robotics 99,
DOI: 10.1007/978-3-319-04123-0_4, © Springer International Publishing Switzerland 2014

focused on skin stretch in haptic devices. Authors in [35] developed a haptic device that displays grasping and weight sensations based on cutaneous stimuli, in which the motions of two motors create controlled skin stretch on a fingertip depending on the teleoperative objects. Also, authors in [36] developed a device for studying haptic behavior by means of skin deformation of the human fingertip. Thus, there is also an increasing need for modeling skin deformation of the human fingertip during contact with an object, especially during the pre-slide or sliding phases. Konyo [37] attempted to model a human fingertip by using a simple mass-spring-damper model to investigate stick-slip events. However, the result was inadequate as the structure of a real human fingertip is very different. In the virtual world of a haptic system, it has been shown that in order to assure a successful tactile exploration, it is indispensable to create a significant *sense* of realness on the user [38]. Especially for soft fingertips, at low velocity from starting sliding state, on the contact area there are localized/partial movements of contacting points, which featured as a pre-slide regime. Several friction models such as Dahl or LuGre [39] are successful at representation of pre-slide phase of soft objects as an increase of friction force over movement of the object. Nonetheless, the details of partial movements on the contact area during the stick phase are impossible to be attained.

In this thesis, we employed our proposed Beam Bundle Model (BBM) (see previous chapters) in the attempted modeling of a human fingertip. In order to implement this, we utilized magnetic resonance imaging (MRI) of a subject's fingertip to construct a mathematical model of the structure of a fingertip. We first characterize pre-sliding regime on a real human finger in order to find out which factors is the most important in term of haptic display. We succeeded in creating a representation of localized displacements (also referred to as local skin stretch) during the pre-slide phase of the fingertip, which is considered crucial to assessing stick/slip events on the contact surface during contact with the outside world. The results of this research can be utilized in further studies in haptic sensation, and to develop sensors for the detection of slippage.

4.2 Human Fingertip Frictional Characteristic

Any of current robotic fingertips is far from human fingertip, in term of structure and function. Human fingertip basically includes three main layers: epidermis, dermis, and bone. In addition, there are a high density of blood vessels and lymphatics forming a network in the tissue, as well as mechanoreceptor system. Thus, it is a complicatedly inhomogeneous structure, in which the soft tissue deform and recover continuously depending on external load. As a result, it is necessary to characterize human fingertip's slip action to assess what kind of phenomena dominated; and in order to introduce these realistic issues to the proposed model.

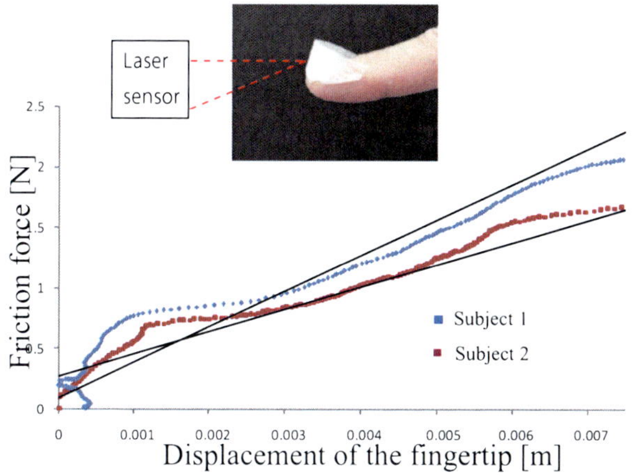

Fig. 4.1 Linear relation between friction and movement of the fingertip

We set up an experiment for characterization of human fingertip, in which subjects were asked to push and slide their fingertips on a rigid, transparent surface. The translational movement of the finger was tracked by a laser range sensor (LB-01, Keyence, Japan) with resolution of 0.1 mm, while three components of force applied to the surface was recorded by a 3-DOF loadcell (USL6-H5-50N-C, Tekgihan, Japan). Moreover, we employed a pressure distribution sensor (ISCAN-50, Nitta, Japan) to assess normal force distribution when pushing the finger. In order to get *detail* deformations of the contact area of the fingertip during sliding, a high speed camera was attained to fulfill this mission. A display showing current values of components of force, and pressure distribution was setup for subjects to observe so that they can adjust applied loads as required. As mentioned in Section I, we focused on pre-slide regime of the human fingertip.

4.2.1 Relation of Friction Force and the Finger's Movement during Stick Phase

Some friction models such as Dahl or LuGre have mentioned about *spring-like* relation between friction force and displacement of object, especially soft objects, during stick phase. We have investigated this characteristics on human fingertip by asking subjects to start to slide their finger slowly enough so that its movement can be captured by laser displacement sensor, while loadcell measures friction force acting on it. A light-weight shield was attached on the fingertip's nail for assurance of correct reflection of the laser beam from the displacement sensor (see the inset picture in Fig. 4.1). The subject might change inclination angle of contact or contact depth. Plots shown in Fig. 4.1 show roughly linear relation between the friction force

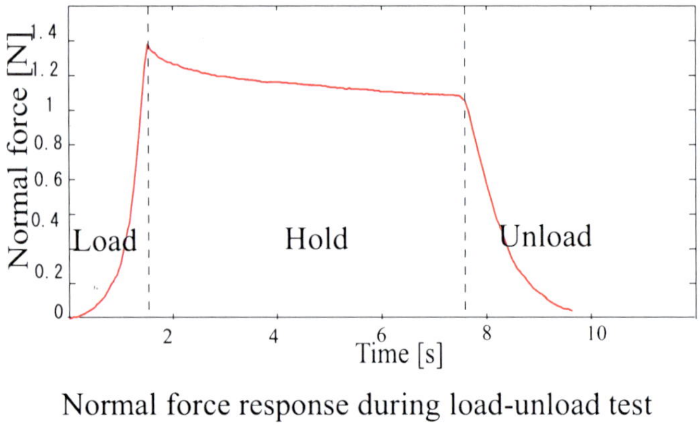

Fig. 4.2 Experimental setup and force-related results

and displacement of the fingertip. This indicates that there exists a stick phase prior to the gross slide of the fingertip, in which even the fingertip already gives a way, the contact area still sticks, causing the increase of the friction mentioned in [39]. As a result, this phenomenon also occurs in human fingertip, and the proposed model must capture this realistic issue into its form. However, the sole model of a linear spring as proposed in [38] is not sufficient to represent this phenomenon, since human fingertip's friction also suffers a high hysteresis over sliding velocity.

4.2.2 Relaxation of Normal Force

Fig. 4.2.2 (a) shows a normal force distribution on the contact area when a fingertip was pushed at a specific contact depth. Differing to that of typical robotic fingertips such as cylindrical or hemispherical ones, this distribution is unsymmetrical, and unpredictable if only based on the boundary geometrical shape. Moreover, the response of the normal force during loading, holding, and releasing phases varies significantly when the rate (of loading/unloading) changes as illustrated in Fig. 4.4(b). There also exists a relaxation when the fingertip is holding its position at a specific contact depth. As a result, it is insufficient to introduce sole elasticity to the model of the human fingertip as mentioned in [38] and [37].

4.2.3 Localized Displacement during Stick Phase

There have been many experimental research figuring out partial movements on the contact area during the pre-slide phase (*i.e.* stick phase as aforementioned) of soft robotic fingertip [12]. It is expected that this phenomenon also occurs on human fingertip, thanks to its soft dermis and curvature shape. In

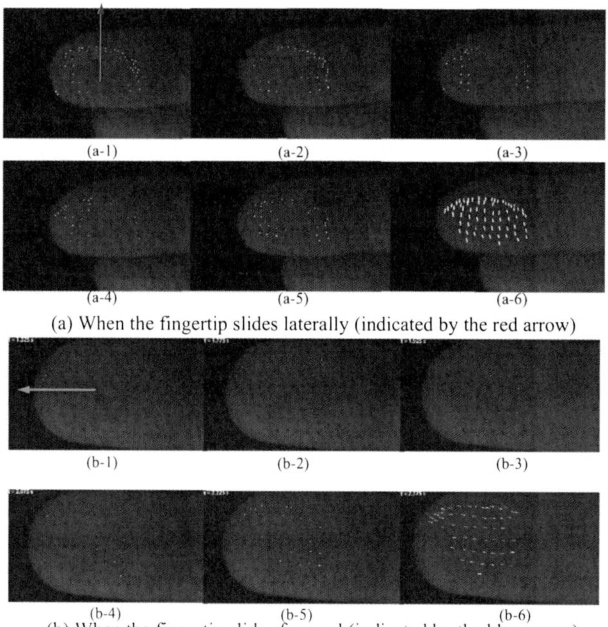

(a) When the fingertip slides laterally (indicated by the red arrow)

(b) When the fingertip slides forward (indicated by the blue arrow)

Fig. 4.3 Movements on the contact area during the stick phase. White dots/bars indicated the slipped points on the contact pad.

this experiment, we attempted to verify this phenomenon by observing movements of dots marked on human fingertip's skin on the contact area by the high speed camera. We deposited black dots onto subjects' fingertip, then asked them to slide the finger gradually from zero along different directions, attempting not to change velocity and normal force. We then employed optical tracking method that mentioned in [42] to track contacting points. Fig. 4.3 presents tracking results of dots on the contact area, in which movements of slid dots are indicated by white bars over time during pre-slide regime; revealing that there exist *localized displacements* of contacting dots in the stick phase. This displacements start from boundary areas of the contact pad, then propagate to the inner area, resulting that the slip phase only occurs when these partial movements dominate all over the contact pad. It is also clear that partial movement propagation differs when the direction of slide changes. Similar results also can be found in [27]. Up to date, no analytical model could tackle this issue in its closed form, since transition between stick-slip states of each contact point has inherited discontinuity. As a result, a model in which generalized displacements of contact point can be derived is necessary with the introduction of stick-slip transition.

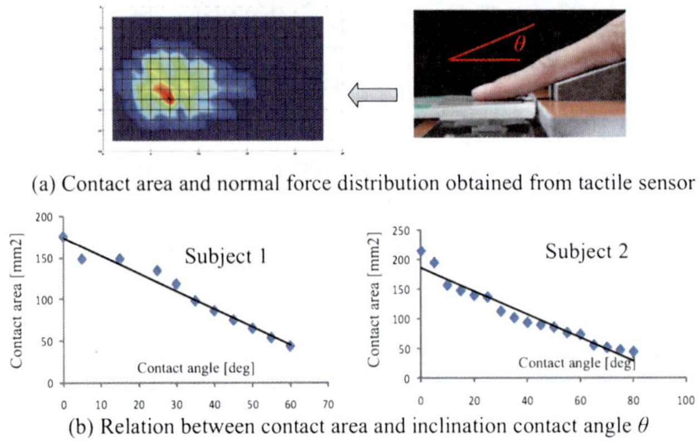

Fig. 4.4 Linear decrease of contact area over change of inclination contact angle during rotation action

4.2.4 Relation of Contact Angle and Contact Area

Some contact models like Hertizan [25] or power model [7] dealing with contact area of soft objects during normal contact. However, to date, no model has addressed role of contact area during relative slide. In this experiment, we asked the subject to change contact depth, *i.e.* contact area varies, during sliding. It is obvious that responses of resulted friction change correspondingly since normal force differs caused by change of contact depth. The most interesting result comes from trials in which subjects attempted to keep normal force unchanged, then rotated the finger to change inclination contact angle. We found that the contact area reduces linearly to the increase of contact angle (Fig. 4.4). This is much different from typical robotic soft fingertips, such as cylindrical or hemispherical ones, in which the contact area remains unchanged during rotation. Consequently, in order to fully describe inherent characteristics of human fingertip in the model, issues mentioned above, more or less, must be included.

4.3 Related Works

4.3.1 Fingertip Modeling in Robotics

The modeling of a biomimetic structured fingertip has been an interesting subject in the field of robotics for years. The process starts with an homogeneous soft-fingertip model, in which the fingertips have simple geometrical shapes, such as cylindrical or hemispherical ones. [19] proposed a linear finite element method (FEM) model of a simple fingertip considering the positions of strain gauges in order investigate the responses of these sensing systems during the grasping an object. [7] conducted a nonlinear FE analysis to study the contact mechanics of a hemispherical fingertip. These researchers mostly focused on normal contact that causes a pressure distribution profile over the contact zone in order to verify the power-law theory in material mechanics. We also recently introduced a 3-D model of an elastomeric hemispherical fingertip in the context of sliding motion in order to optimize the location of a micro sensor used to sense multiple states of contact with an exterior environment (see previous chapters).

To further understand human fingertip functions so they can be mimicked in a robotic fingertip, a uniform model is not completely adequate; thus non-homogeneous models having multiple structures, comprising tissue, bones, and nails, have been constructed recently. [40] have proposed an FE-based 2-D model of a soft fingertip with a simple bone structure. Simulation results show that the addition of a bone in many configurations contributes significantly to the force distribution, as well as to responses from mechanoreceptors. More complicated models of the human fingertip with accurate geometries generated from CT images were proposed in [41] for the study of sole deformation of the fingertip on a flat plate. More recently, [27] constructed 3-D FE models of the fingertips of 50 subjects working from magnetic resonant images and showed variations among subjects in terms of contact pressure during contact. The above studies, however, only addressed a static model for predicting stress-strain when making contact; and none of them could reappear dynamic changes in stress-strain in a time domain. For research related to the study mechanism of the work of a mechanoreceptor for application in a haptic sensation, term dynamics are considered important for an assessment of this issue in its entirety.

4.3.2 Modeling of the Sliding Motion of a Fingertip

The sliding action is a crucial motion for a fingertip to perform so that it can feel an object and act accordingly in order to prevent it in grasping or manipulation. Moreover, it is also important in term of tactile perception, in which a human/robot needs to slide a finger over a surface so that its textural characteristics can be detected by mechanoreceptors underneath the skin and decoded by the brain/controller. For that reason, some researchers have addressed sliding motion in their work. [12] built a closed form for modeling manipulation with sliding fingers. However, this research was limited to quasi-static simulation and homogeneous fingertips, and could not help in assessing details of slippage on the contact area. Konyo [37] has investigated the human fingertip under vibrotactile stimuli in order to represent force in a haptic display system. A dynamic model of the human fingertip was proposed to study stick/slip events on the contact surface while subject to external high-frequency stimuli. However, this model simply comprised a mass connected with a spring and a damper that were arranged in parallel, which is insufficient for a complexly structured human fingertip. Recently, the authors in [43] proposed a complete FE model of the human fingertip using a dynamic approach. However, the sliding action has not been addressed fully and there is no friction force. Moreover, the model takes a great deal of time to implement a simulation trial, and is considered impractical for use in a real-time haptic system. Recently, Nahvi *et al.* [38] has introduced a friction display system for virtual environment with a simplified model of a virtual spring during transition from slip to stick, and vice versa. While this model can extract human fingertip's characteristics such as switching between stick and slip, oscillation in slip phase; it cannot clarify how partial slips occurs on the contact area during stick phase that is, as stated by authors, crucial to assess slip perception.

4.3.3 Beam Bundle Model of a Soft Fingertip

Previously, we proposed the BBM for dynamically modeling the sliding motion of cylindrical and hemispherical soft fingertips. In this model, the soft fingertip is necessarily elastic and homogeneous with a predetermined geometrical shape (see Fig.4.5). First, in order to model the deformation of the fingertip under normal and tangential loads, we filled the fingertip with *virtual* elastic beams that are compressible and bendable, which means the deformation of the fingertip can be calculated based on the deformations of the beams. Second, the contact area was meshed with a Kelvin-Voigt model according to finite element theory. The free ends of beams were placed at nodes on the contact area, thus the mutual interactions of the beams were considered to take place solely on the contact surface. Detailed derivations of the motion equations can be referred to in Part I of this book. Using this model, we were able to simulate sliding motion under a given speed, and observe the responses for normal force and friction force. Moreover, the simulation also produced localized displacement phenomenon during the pre-slide phase, helping us to assess how and when the slippage occurred on the contact surface. This is considered important to understand slip of a soft object, as well as being crucial to tactile sensation. Having this tool ready, we will explain how to use it to model the human fingertip in the following sections.

4.4 Constructing a Fingertip Model Based on MR Images

In order to accurately measure the 3-D internal and external geometries of a human fingertip, a 3 Tesla (3T) MRI system, SIGNA HDXT (GE Healthcare, Waukesha, WI), was used in our experiments. Cross-sectional MR images of an index finger were collected from a 25-year-old adult male, having no history of finger disease. A set of 32 images with a $120\times120\,\mathrm{mm}^2$ field

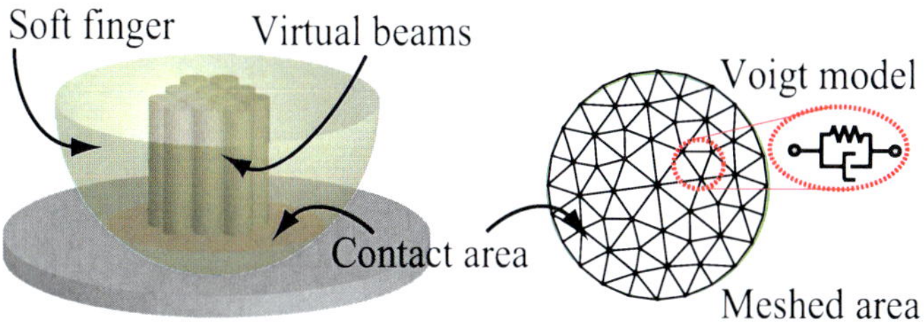

Fig. 4.5 Beam Bundle Model

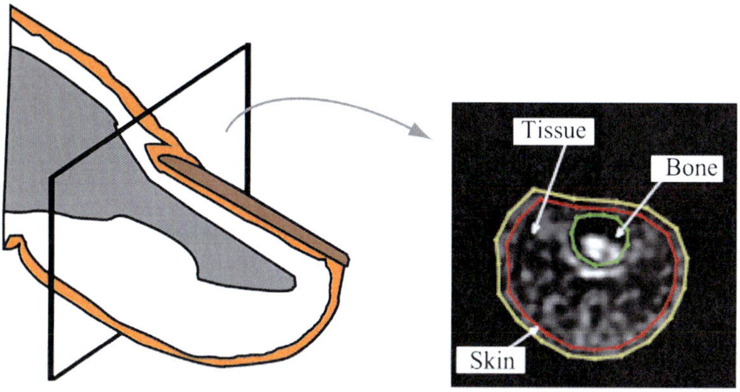

Fig. 4.6 MR images of consecutive cross-sectional layers of a human fingertip, showing the distribution of skin, phalanx, and nail

of view (FOV), 1.2 mm slice thickness and 512×512 matrices (pixel size of $0.23 \times 0.23 \times 1.2\,\text{mm}^3$) were acquired using a 3-D fast gradient echo (fgre) sequence with a 9.8 ms repetition time (TR), 4.2 ms echo time (TE) and 30 flip angle. These images were obtained representing volume of the fingertip in terms of consecutive cross-sectional layers. In each image, we can observe the distribution of layers such as skin, tissue, bone, and nail position as illustrated in Fig. 4.6.

To introduce the BBM for use in modeling the human fingertip, it is necessary to assess the exact distribution of the inner layers in order to fill the inner volume with virtual beams properly. We utilized image-processing functions in OpenCV to extract the boundaries of skin, bone, and the position of the nail. Note that as the MR signal is mainly derived from the protons of water molecules in body tissues, we were not able to extract the shape of the nail, which is a rigid material, only the boundary on which the nail is situated. Each boundary was formed by a group of points, that were later interpolated into a curve (see Fig. 4.7(a)). As a result, for each image, we were able to collect four fitted curves, including for skin, lower bone, upper

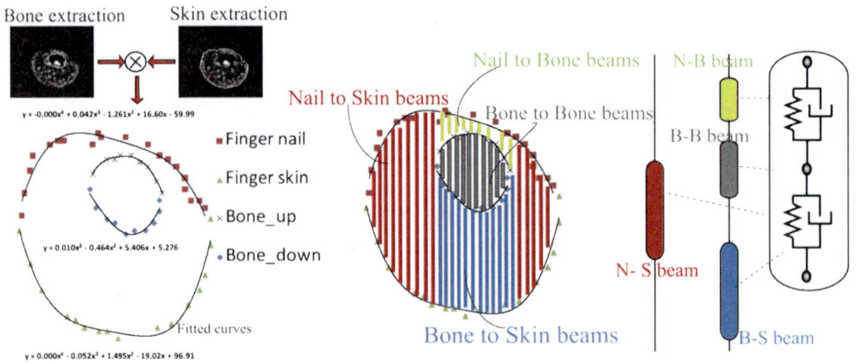

(a) Structure extraction (b) Groups of virtual beams inside the fingertip (c) Beams with generalized Voigt element

Fig. 4.7 Structure of human fingertip from MRI

bone, and nail as illustrated in Fig. 4.7(a). By repeating this process for all images, the 3-D geometrical shape of the fingertip could be constructed for the introduction of the BBM.

4.5 Beam Bundle Model of Human Fingertip

After obtaining boundary curvatures of the human fingertip structure, we need to fill in the remained volume with *virtual beams*. These beams have two ends constrained by an upper bound (nail) and a lower bound (skin). Each virtual beam possesses cylindrical shape with geometrical property, such as height, predetermined based on known fitted curves of nail, bone, and skin. Since layers of MR images have a pitch of 1.2 mm along the volume of the fingertip as mentioned in the above section, and in order to fill in beams to fit nicely to the fingertip's volume, each beam has a cross-sectional area's diameter of 1.2 mm. As a result, distance between basic axes of two neighboring beams is also specified as 1.2 mm.

In order to reflect relaxation mentioned in Section 4.2.2, we assigned each beam with a *generalized Voigt* model that includes two Voigt elements, each element has viscoelasticity characterized by a spring and a damper connected in parallel. After filling in with virtual beams, as illustrated in Fig. 4.7(b), there are two main groups of beams, including beams attach to nail and skin (N-S) and beams attach to nail through bone (N-B-S). Beams in the latter group are constituted by three separated beams connected in series: N-B, B-B, and B-S as shown in Fig. 4.7(c). As a result, virtual beams are not necessarily homogeneous as in the previous chapters. By using this proposal of inhomogeneous beams, we could fill in *any* complicatedly-structured soft fingertip.

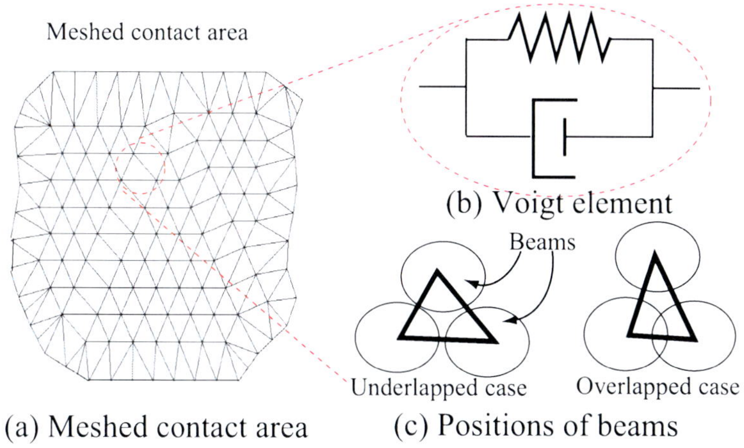

Fig. 4.8 Meshed contact area

Supposed that the fingertip was pushed vertically onto a rigid flat surface, causing a set of contact boundary points on skin. We then meshed the contact area (or skin) utilizing finite element analysis, covered the contact area with a set of triangles (Fig. 4.8(a)); in which stress-strain relationship of each triangle is represented by a Voigt element that includes a spring and a damper connected in parallel (Fig. 4.8(b)). Each triangle has three nodes attached to three corresponding free-ends of three beams that belong to either N-S or B-S group (see Fig. 4.8(c)). There exists two types of contribution of beams on one triangle: partly covering, and superposition. The former occurs more often then the latter case, causing coverage ratio of entire contact area is always less then 100%. The coverage varies depending on how coarse or fine we meshed the contact area, and the maximum ratio that has been reached with less superposition is 96%. As a result, we can flexibly change the coverage ratio by varying beams' geometrical shape to adapt to specific simulation.

Thus, beams are constrained on the contact area on the skin; and the movement of beams's free-ends would be helpful to assess stick/slip events on the contact area during sliding motion. We have named this model as Beam Bundle Model (BBM).

4.5.1 Avoiding Re-meshing the Contact Area at Different Contact Depths

When a fingertip slides on a surface or several fingertips grasp an object, it is common that contact depth may vary to adapt to the outside environment. In this scenario, the contact area meshed by finite elements must be also adjusted to a different geometrical characteristic compared to its initial contact. This causes the number of virtual beams and their geometry alter significantly. This is inconvenient in term of programming and control, since it produces discontinuity during change of contact depth. We thus propose a method that permits *one* time of meshing, with no remeshing during the contact depth varies.

Note that each node on the meshed contact area corresponds to one beam that located exactly on this node. Thus, there exists two groups of beam: contact beams positioned inside the contact, and non-contact beams that are outside the contact area. When the contact depth varies, two following conditions must be satisfied:

1. Contact beams that exists at different contact depths must be unchanged. Their corresponding nodes, thus, unchanged.
2. Non-contact beams that become contact beams at the updated contact depth must be assigned a *pre-determined* node on the contact pad.

We, therefore, *only* mesh the contact area at the possibly *maximum* contact depth, says d_{max}. At this contact depth, the contact area is meshed with a set of N nodes $\{n_i, i = 1 : N\}$, and corresponding N beams $\{b_i\}$. These nodes and beams are unchanged throughout different contact depths. Given a specific contact depth $d < d_{max}$, each beam is checked whether it lies inside or outside the new contact area. Depending on its position, its corresponding node will be marked as contact node n_j^c ($j = 1 : N_c$, N_c is the total number of contact node), or non-contact node n_k^{nc} ($j = 1 : N_{nc}$, N_{nc} is the total number of non-contact node, and $N_c + N_{nc} = N$). As a result, on the meshed area, there are two groups of nodes $\{n_j^c\}$ and $\{n_j^{nc}\}$. We supposed that non-contact nodes were *virtual* nodes, which only exist to assure that the previously meshed area's geometrical characteristics unchanged but have no affection to the deformation of the entire area, as well as to the movements of contact nodes. In order to satisfy this condition, for one triangle on the meshed area constituted by three nodes, if at least one of three vertices is virtual, the triangle's stress-strain relation will be eliminated. In details, the stress-strain relationship in a triangle can be converted into a relationship between a set of forces $\mathbf{f}$ applied to nodes and their displacements $\mathbf{u}$. This set of forces of a Voigt element is calculated as following well-known relationship:

$$\mathbf{f} = \mathbf{J}_\lambda(\lambda^{ela}\mathbf{u} + \lambda^{vis}\dot{\mathbf{u}}) + \mathbf{J}_\mu(\mu^{ela}\mathbf{u} + \mu^{vis}\dot{\mathbf{u}}), \tag{4.1}$$

where $\mathbf{u}$ and $\dot{\mathbf{u}}$ are displacement and velocity vectors of nodal points, $\mathbf{J}_\lambda$ and $\mathbf{J}_\mu$ are connection matrices that depend solely on geometric coordinates of nodes, $\lambda^{ela,vis}$ and $\mu^{ela,vis}$ are Lame's constants that are described by Young's modulus E, Poisson's ratio γ, and viscous modulus c. As a result, to vanish interaction of $\{n_j^c\}$ to $\{n_j^{nc}\}$, we simply set Lame's constants of a triangle that owns non-contact nodes, to zero temporarily for this contact depth. If any of $\{n_j^{nc}\}$ becomes $\{n_j^c\}$, its corresponding triangle's Lame's constants will recover their original values. As a result, simulations always run on one

Table 4.1 Physical Parameters for the Human Fingertip ([43])

	Tissue	Skin	Bone
E[Pa]	3.4×10^4	1.36×10^5	1.5×10^9
c[Pa.s]	Not used	10	Not used

meshed contact area during change of contact depth, taking into account only real contact nodes and giving virtual nodes none-stress/strain relation.

4.5.2 Derivation of Motion Equation

In this section, we summarize some important equations for describing dynamic movement of the fingertip itself, as well as the contact area. Derivations of the motion equations for the whole fingertip are similar to the hemispherical phase, which can be found in Chapter 3 of this book.

Suppose that the fingertip is pushed vertically onto a flat surface, then slid horizontally with a constant speed. The virtual beams are deformed correspondingly, including normal and bending deformation. There are several assumptions made to simplify the calculations:

1. Interactions between neighboring beams only occur among their free ends *on* the contact zone.
2. Only beams whose free ends are acting on the contact surface are considered. Beams outside the contact surface are deemed irrelevant to the sliding motion of the fingertip.
3. Nail and bone (phalanx) are completely rigid and suffer no deformation during the sliding motion of the fingertip.
4. The area of the nail is always larger than the contact area.

The first and the second assumptions are important for drastically reducing calculation costs. This is because of the fact that the tissue volume of a fingertip deforms quite uniformly, while skin on the contact area can be stretched locally differently, the first assumption is acceptable. Since superposition of the beams is small and irrelevant to the precision of the model, beams deformed by a pushing action do not affect beams positioned outside the boundary of the contact area, The second assumption is sufficient for modeling the deformation of an entire fingertip. If we note the Young's modulus of volumes in Table 6.3, that of bone and nail is much bigger than that of tissue or skin. Thus the third assumption is also acceptable for calculation without degrading the precision of the model. This assumption also allows us to neglect the deformation of the N-B group of beams as they are positioned between two undeformable layers of nail and bone. As a result, we only consider the deformation of the two remaining beam groups: N-S and B-S (see Fig. 4.7(b)). The last assumption is rather more of a reality as most of a human fingertip has a nail that covers almost the entire upper part of the fingertip. This assumption allows us to eliminate beams on the contact area that make contact neither with bone nor nail, but are pinned to the skin. By accepting the above assumptions, we are able to proceed with the force calculations and motion equations.

During Loading/Unloading

In this section, we derive general equation for obtaining normal deformation of one inhomogeneous beam constituted by set of Voigt models connected in serial, which is referred to as generalized Voigt model.

Supposing that one beam consists of n Voigt elements, let us define ε be the extensional strain of the beam while ε_i be the extensional strains of the i-th element. Extensions ε and ε_i through ε_{n-1} are independent state variables, and ε_n can be described dependently by other variables as follows:

$$\varepsilon_n = \varepsilon - \varepsilon_1 - \ldots - \varepsilon_{n-1}. \tag{4.2}$$

Let E_i and c_i be the elastic and viscous moduli of the i-th Voigt element. Noting that stresses generated by individual elements are equal to each other, we have:

$$\sigma = E_1\varepsilon_1 + c_1\dot{\varepsilon_1} = E_2\varepsilon_2 + c_2\dot{\varepsilon_2}\ldots = E_n\varepsilon_n + c_n\dot{\varepsilon}, \tag{4.3}$$

where σ indicates the generated stress by the deformation of the beam. Dividing the above equations by c_1 through c_n and summing up all equations yields:

$$\sigma = \frac{\left(\frac{E_1}{c_1}\varepsilon_1 + \ldots + \frac{E_{n-1}}{c_{n-1}}\varepsilon_{n-1}\right) + \frac{E_n}{c_n}(\varepsilon - \varepsilon_1 - \ldots\varepsilon_{n-1}) + \dot{\varepsilon}}{\frac{1}{c_1} + \frac{1}{c_2} + \ldots\frac{1}{c_n}}. \tag{4.4}$$

Note that using the above equation, stress σ can be computed from state variables ε_1 through ε_{n-1} and $\dot{\varepsilon}$. State variables satisfy the following differential equations:

$$\dot{\varepsilon_1} = -\frac{E_1}{c_1}\varepsilon_1 + \frac{1}{c_1}\sigma, \ldots, \dot{\varepsilon}_{n-1} = -\frac{E_{n-1}}{c_{n-1}}\varepsilon_{n-1} + \frac{1}{c_{n-1}}\sigma. \tag{4.5}$$

Consequently, we can construct dynamic equation for calculating deformations of elements in the inhomogeneous beam. We repeat this for all virtual beams in the fingertip. This derivation is also exploited to model any inhomogeneous soft objects other than fingertips.

Sliding Motion

When external tangential force F_t starts to activate, the fingertip has not yet slid. The contact surface still sticks to the plane, causing the fingertip

Table 4.2 Simulation Parameters

Sliding speed [mm/s]	Friction coefficient	Direction of slide
2-5	0.6-2.0	X+, X-, Y+, Y-

to deform. At this time, all the contacting cantilevers are bent at the free ends under the same bending strain δs. This bending strain is calculated as originated in [34]:

$$\delta s = \frac{3\mu F_n}{16R}\frac{2-\nu}{G}\{1-(1-\Phi)^{2/3}\}, \tag{4.6}$$

where $\Phi = F_t/\mu F_n$ is the tangential force coefficient, R is the radius of the fingertip, μ is the friction coefficient, ν is the Poisson's ratio, and G is the shear elasticity modulus. By assessing the bending strain in Eq. 4.6, the bending force f_b^i acting on the free end of the cantilever, with length l^i and cross-sectional area radius r^i, is calculated in the following equation:

$$f_b^i = b^i \delta s = \frac{3EI}{(l^i)^3}\delta s = \frac{3E\pi(r^i)^4}{(l^i)^3}\delta s. \tag{4.7}$$

For a FE-based meshed contact area, a viscoelastic force exists at each contact node, which is calculated based on the node's geometrical characteristic in the context of interaction with neighboring nodes as follows:

$$\mathbf{F}_{ve} = (\lambda^{ela}\mathbf{J}_\lambda + \mu^{ela}\mathbf{J}_\mu)\mathbf{u}_N + (\lambda^{vis}\mathbf{J}_\lambda + \mu^{vis}\mathbf{J}_\mu)\mathbf{v}_N, \tag{4.8}$$

or:

$$\mathbf{f}_{ve} = \mathbf{K}_{ela}\mathbf{u}_N + \mathbf{K}_{vis}\dot{\mathbf{u}}_N, \tag{4.9}$$

where λ^{ela} and μ^{ela} are the elastic Lame constants; λ^{vis} and μ^{vis} are the viscous Lame constants. The connection matrices can be described as $\mathbf{J}_\lambda$, and $\mathbf{J}_\mu$ and can be obtained by synthesizing a set of triangles of partial-connection matrices.

We also introduce a friction law into each node on the contact surface. Its value changes based on its contact state: stick or slip, and it is calculated as follows (see Chapter 3):

$$\mathbf{f}_{fr}^i = \begin{cases} \mathbf{f}_{ve}^i + \mathbf{f}_b^i, & \mathbf{f}_{ve}^i + \mathbf{f}_b^i < \mu\mathbf{f}_n^i \Rightarrow Stick \Rightarrow \mathbf{A}^i = \mathbf{A}_1^i \\ \mu\mathbf{f}_n^i, & \mathbf{f}_{ve}^i + \mathbf{f}_b^i \geq \mu\mathbf{f}_n^i \Rightarrow Slip \Rightarrow \mathbf{A}^i = \mathbf{A}_0^i \end{cases} \tag{4.10}$$

Finally, by utilizing the Lagrangian formula and the constraint stabilization method (CSM), we were able to construct motion equations for all contacting nodes on the contact area as follows:

$$\begin{bmatrix} \mathbf{I} & \mathbf{0} & \mathbf{0} \\ \mathbf{0} & \mathbf{M} & -\mathbf{A} \\ \mathbf{0} & -\mathbf{A}^T & \mathbf{0} \end{bmatrix} \begin{bmatrix} \dot{\mathbf{u}}_N \\ \dot{\mathbf{v}}_N \\ \lambda \end{bmatrix} = \begin{bmatrix} \mathbf{v}_N \\ -\mathbf{K}_{ela}\mathbf{u}_N - \mathbf{K}_{vis}\mathbf{v}_N + \mathbf{F} \\ \mathbf{A}^T(2\omega\mathbf{v}_N + \omega^2\mathbf{u}_N) \end{bmatrix}. \tag{4.11}$$

This equation is linear and solvable since the matrix is regular, implying that we can compute $\dot{\mathbf{u}}_N$, and $\dot{\mathbf{v}}_N$, which are the displacements and velocities of the contacting nodes, respectively. Details of this derivation can be found in Chapter 3 of this book.

4.6 Simulating Human Fingertip Sliding

In this simulation, the fingertip was given a vertical push at a predetermined contact depth and a tangential movement with constant velocity. The parameters used for the simulation are summarized in Tables 6.3 and 6.4.

4.6.1 Force-Related Results

Figure 4.9 shows the normal force distribution when the fingertip was pushed at a contact depth of 2 mm. We can see that the maximum force area is around the tip, which is close to the nail, while normal forces with smaller values are distributed mostly near the boundary of the contact surface. This result is similar to the calculation of [27] using the FEA model. As a result, it is noticeable that although the outer shape of a human fingertip is generally symmetrical and even, the force distribution is not produced so neatly as in the case of a hemispherical homogeneous fingertip (see Chapter 3). It is the inner distribution of the distal phalanx that causes remarkably different geometrical shapes, as well as nonuniform deformations of beams over the contact area.

Figure 4.10 plots the response of the friction force during the stick-to-slip phase under constant speed. Similar to the hemispherical case in Chapter 3, two phases featuring a change of friction force are easily identified. The stick phase, which represents the pre-slide phase, when the fingertip starts to move but the contact area still sticks to the surface, *i.e.* complete movement of the fingertip has not occurred yet, and the friction force increases remarkably. The slide phase, which indicates the total movement of the contact area, and the friction force goes to unchanged value. There is a smooth transaction between the two phases. This phenomenon is similar to that for the hemispherical soft fingertip discussed in Chapter 3. The friction force changed drastically when we varied some parameters, such as the sliding speed and contact depth. As plotted in Fig. 4.10(a), the responses of friction force alter over time through a wide range of sliding speeds, indicating that the stick phase occurs in a small amount of time when the speed increases. Contact

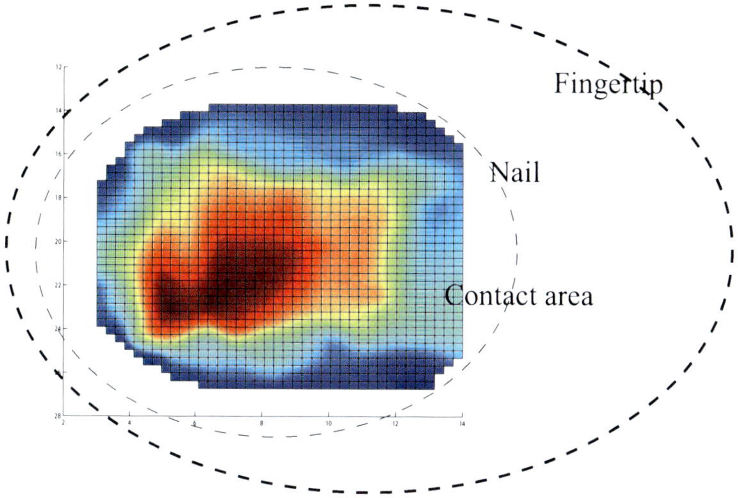

Fig. 4.9 Normal force distribution on the contact surface in context of fingertip

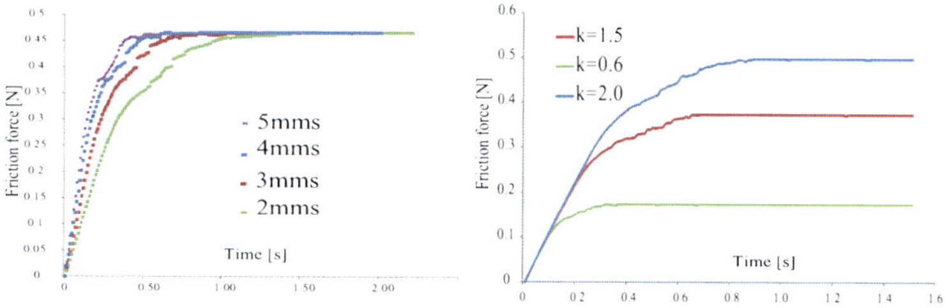

(a) Friction when sliding speed changes (b) Friction when friction coefficient changes

Fig. 4.10 Friction force during stick-to-slide phase

depth also has a remarkable affect on friction as exemplified in Fig. 4.10(b). For a greater contact depth, not only does friction augment its value, but the stick phase also elongates over time. This is due to the fact the normal force increases in advance when the fingertip is pushed more deeply. We also changed the friction coefficient over the range 0.6-2.0, which is typical for human fingertips. As illustrated in Fig. 4.11(a), the model works well for friction coefficients larger than 1, showing that the magnitude of the friction force and the stick phase's period both increase for fingertips having higher coefficients.

Finally, we investigated the dependence of the friction on the fineness of the meshed contact area. Figure 4.11(b) shows that for a coarse mesh, the friction force has a smaller value in terms of magnitude, but maintains a

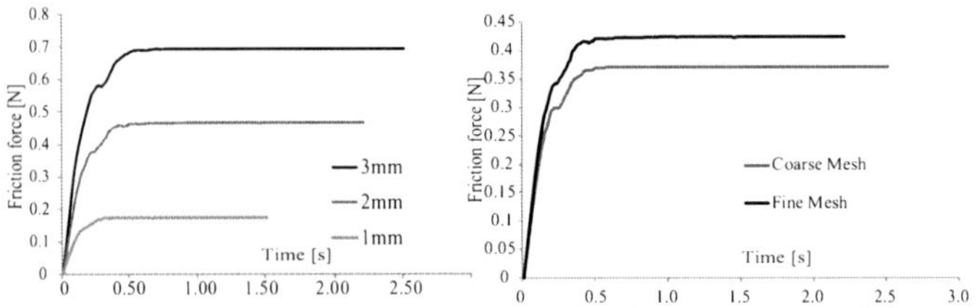

(a) Friction when contact depth changes (b) Friction when contact area mesh changes

Fig. 4.11 Friction force during stick-to-slide phase

similar stick phase time. This can be explained based on the coverage ratio of beams over the contact area as mentioned in Section 4.5. In the coarse mesh case, the coverage is 89% compared to 96% for the fine mesh case, resulting in the total normal force of the coarse case being less than for the fine case, resulting in a smaller friction value.

4.6.2 Localized Displacements during the Pre-slide Phase

Of particular note is the successful representation of localized displacements on the contact area during the pre-slide phase in the simulation. There have been several studies on the experimental detection of micro movements on the contact area of the human fingertip during slide action [42] and [50]; however, none of them can explain the underlying theoretical background for this phenomenon. By exploiting BBM for a hemispherical soft fingertip, we have been able to simulate this process. In this research on a human fingertip model, similar results for the phenomenon were assessed clearly. Figure 4.12 shows the distribution of localized movements of contacting nodes on the contact area over moments of time during the pre-slide phase. Bright and hot color zones indicate larger movements than cold and dark color zones. These distribution were taken from a simulation of a sliding trial with a sliding speed of 2 mm/s, a friction coefficient of 0.7 and a contact depth of 2 mm. We can see that the displacement initially occurs near the boundary of the contact area, then propagates gradually along the direction of slide. Another propagation is the spreading of movements from outer areas to inner areas. As a result, this propagation of displacement relies on the normal force distribution, which indicates that small forces are distributed mainly near the boundary rather than the middle part (see Fig. 4.9). Finally, the last area to move immediately prior to the total slippage of the fingertip happens is circled at time 1.6 s in Fig. 4.12 that is coincident to the moment at which the friction starts to be constant in Fig. 4.10(a) (which indicates the start of the gross slide phase of the fingertip).

If we compare the experimental work in [42], in which the authors also pointed out that localized slippage occurs first near the boundary area, then

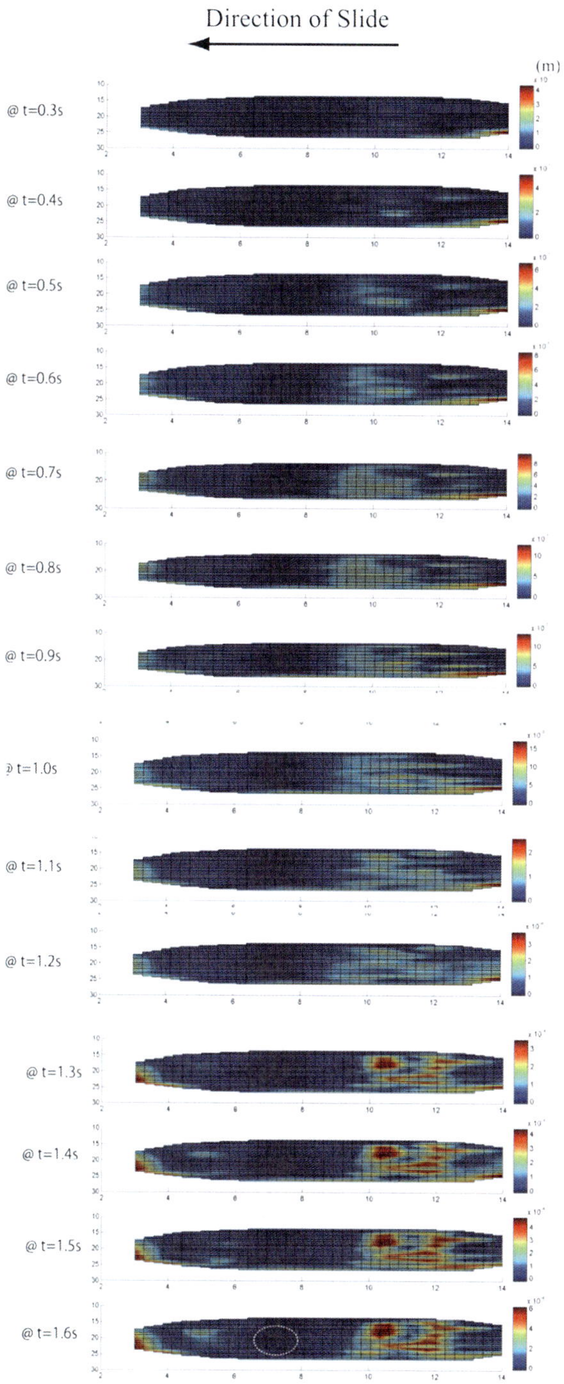

Fig. 4.12 Localized displacement phenomenon

spreads into the inner zone, and the last area to move is near the tip, we find this is similar to our simulation result. This phenomenon, while complex for use in a robotics application, is easily sensed by the high density of mechanoreceptors underneath the skin in a human fingertip. This is the reason humans act so comfortably with respect to the incipient slip of a grasped object as it tends to slip out of the hand. It is a matter of fact that mechanoreceptors do not sense the change of friction force to predict the incipient slip, they are stimulated by local displacement or the stretch of the skin, then tactile signals are sent to the brain for processing, finally control commands are transmitted to the motor system to apply more force to prevent slippage in a real-time scenario. As a result, by modeling and assessing LDP, we are able to understand underlying mechanism of tactile perception of the human fingertip during sliding motion.

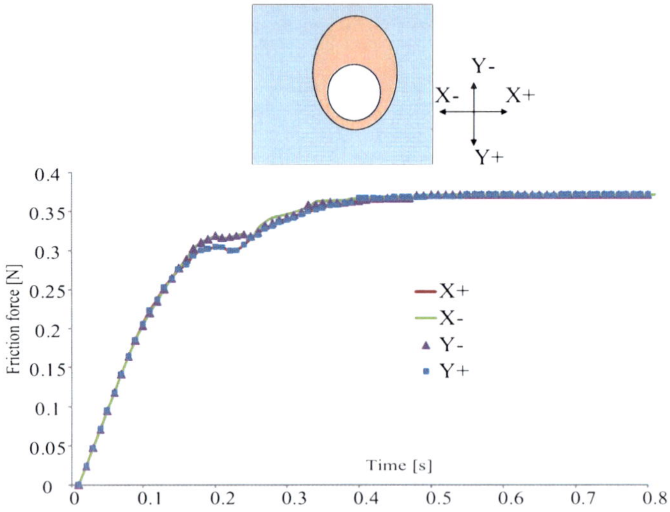

Fig. 4.13 Friction over slide directions

The direction of slide is also investigated in advance. For a hemispherical fingertip, the friction response and the LDP are unchanged, thanks to its symmetry and homogeneousness. However, this conclusion is not correct for the human fingertip due to its complex structure and nonuniform normal force distribution. We conducted simulation trials in which a fingertip was slid along four directions: $X+$, $X-$, $Y+$, and $Y-$ as illustrated in Fig. 4.13. Note that in this graph the friction response along $X+\&X-$ (or $Y+\&Y-$) directions are almost coincident, while a very slight difference is found among frictions along the $X\&Y$ directions. Thus, we cannot use the change in friction as a basis for distinguishing the slide direction. However, interestingly, the LDP has different profiles over various directions. Figure 4.14 shows

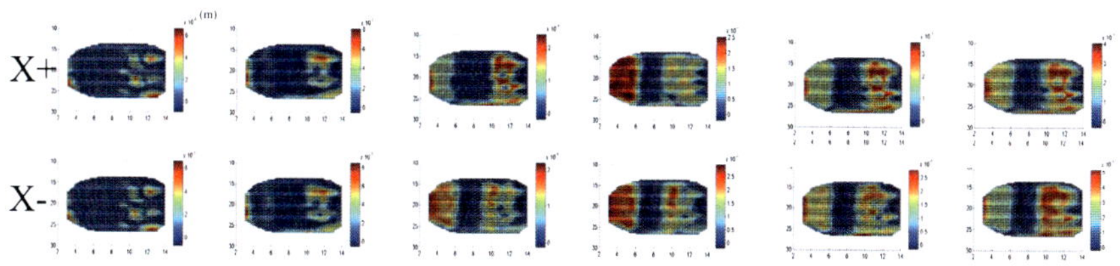

Fig. 4.14 Comparison in terms of local displacement distributions over time when a fingertip slides along the $X+\&X-$ directions

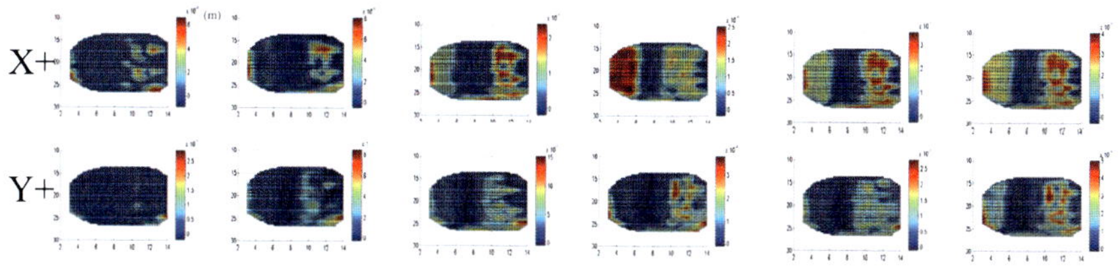

Fig. 4.15 Comparison in term of local displacement distributions over time when a fingertip slides along the $X+\&Y+$ directions

comparisons in term of localized displacement distribution on the contact area at several moments of time during the pre-slide phase along the $X+$ and $X-$ directions. We can see a difference in propagation of local movement on a contact area between two directions of slide, even on the same axis. A similar phenomenon was also perceived between the directions $Y+$ and $Y-$. Even although two opposite directions cause different LDPs, as shown in Fig. 4.14, noticeable differences can be clearly seen when the fingertip slides along the $X\&Y$- axes. In Fig. 4.15, we can easily discriminate distinctive propagation of localized movements on the contact surface along two orthogonal directions. As a result, when the human fingertip slides along different directions, although the friction response does not vary much, the localized displacement phenomenon has a distinguishable profile during the stick-to-slip phase. These different profiles stimulate mechanoreceptors under the skin in different ways, and as a result a human being can sense atypically in various directions.

Consequently, by applying the BBM to the modeling of a human fingertip, we not only assess how and when slippage occurs on the contact area, but we also understand how a human can detect slippage in a timely way based on LDP during the stick-to-slip phase. The way that local slippage erodes the contact area during the pre-slide phase also can suggest a way of developing a sensing system to sense the incipient, and to react to prevent it.

4.6.3 Calculation Cost

All code was implemented in a Microsoft Visual Studio C++ environment, on a common personal computer with an Intel Core Duo 3.0 GHz chipset and 2.0 GB RAM. A typical dynamic simulation trial of 2 sec took about 24 minutes to complete. This is far from the requirements for a real-time application, but faster performance could be implemented in a GPU-equipped computer, and we are currently working toward this. Nevertheless, this computation cost still outperforms typical FE simulation of 3-D fingertip models. A similar analysis using ANSYS software implemented in our previous work (see detailed report in Chapter 5) took six to seven hours to finish. The authors in [43] reported that in their simulation they spent several days running a trial. Our method, therefore, is more promising in terms of a real-time application in the near future.

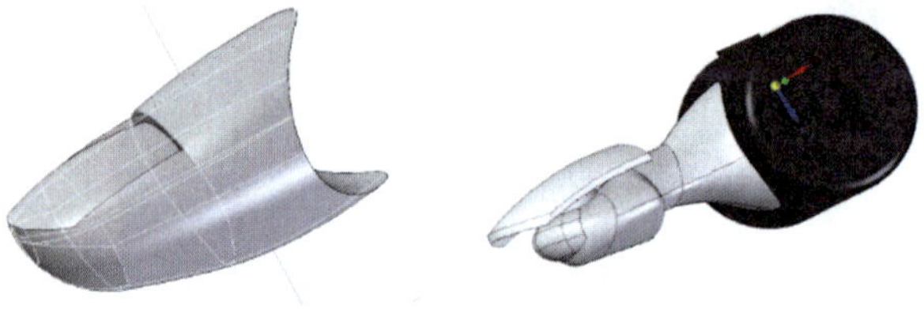

(a) CAD design of the fingertip: soft tissue and bone structure

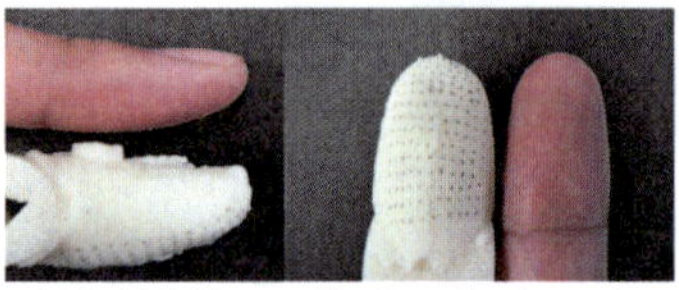

(b) Artificial human-like fingertip after curing

Fig. 4.16 CAD model of the human-like fingertip

4.7 Experimental Validation

We have created an artificial inhomogeneous human-like fingertip based on the MRI data, with surfaces were interpolated and smoothen by CAD program (see Fig. 4.16). The bone structure and nail were made by a 3-D printer, while the soft tissue (softness is similar to that of human tissue) were obtained after curing polyurethane rubber gel in a designed mold. By utilizing this finger, we were able to valid the proposed model precisely. In order to create movements for the fingertip, we attached it onto a 2-DOF linear motorized stage that can provide 2 μm-in-resolution step. The fingertip was made push and slide onto a flat rigid plate, which was then fixed to a table with a 3-DOF loadcell (for measuring three components of force) and a tactile arrayed sensor sheet (for obtaining normal force distribution on the contact area)

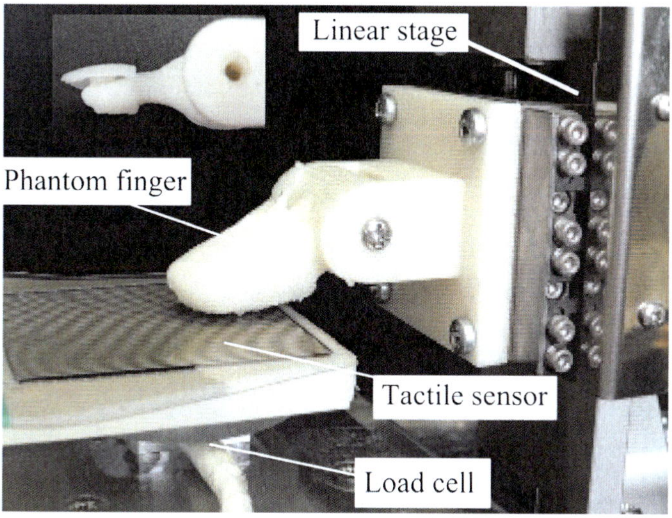

Fig. 4.17 Experimental setup for validation. Inset picture shows artificially made fingertip structure.

(Fig. 4.17). For attaining localized displacements on the contact area, we marked skin of the fingertip with black dots distributed similarly to nodes in the model. These dots were then tracked by a high speed camera during stick-to-slip transition, thus validation of localized displacement phenomenon can be evaluated. With this experimental setups, we can vary conditions of contact, such as angle of contact or direction of slide, thus fully adapts to simulation conditions.

4.7.1 Force Validation

Normal Force Distribution on the Contact Area

The artificial fingertip and the CAD fingertip were made to make a normal contact to the surface at three inclination contact angle: 15 deg, 30 deg, and 45 deg. The total normal force, which is measured by the loadcell, was kept unchanged around 1.7 N through out this validation. In the experiment, the tactile arrayed sensor recorded distributions of normal force, then the data were compared to the simulation results. Fig. 4.18 illustrates results from both simulation and experiment, indicating the agreement among them. We can easily observe the similarity between linear relations of contact area and inclination angle obtained from both simulation and experiment, which also shows similar tendency to result in Section 4.2.4. As a result, with the proposed method, a detail normal distribution can be predicted. We can assess that with the bone structure, formation of force distribution is totally different with that of the homogeneous robotic fingertip. Fig. 4.19 illustrated one

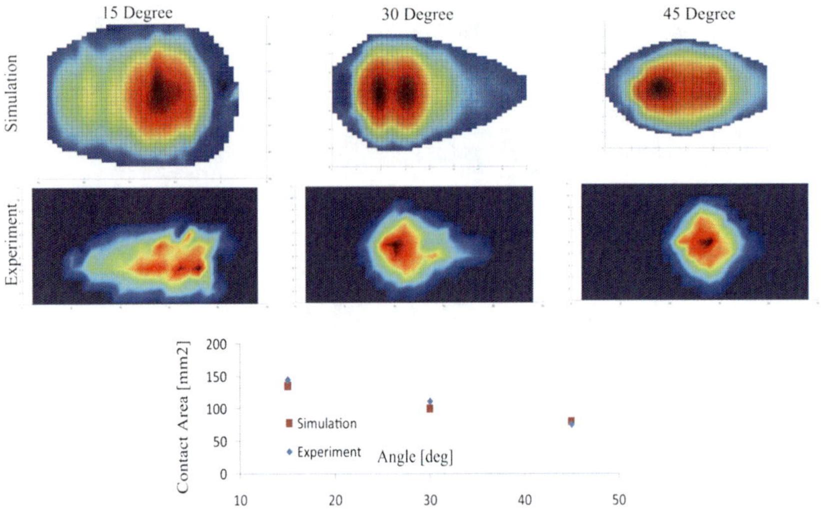

Fig. 4.18 Comparison of normal force distribution from simulation (top) and experiment at different contact angle (bottom). The graph shows the relations of contact area and contact angle obtained from simulation and experiment.

simulated dynamic load/unload test of normal force response. We can observe the exponentially growth/decay of the normal force during loading and unloading phase. Also the relaxation of the normal force when the fingertip was held at the given contact depth, showing a complete agreement to what had been mentioned in Section 4.2.2.

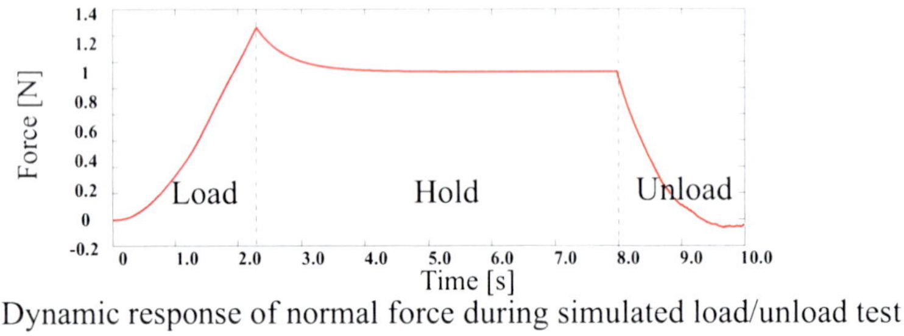

Fig. 4.19 Response of normal force during load/unload test

Friction Response

In this validation, we attempted to compare responses of friction during sliding motion under variant conditions of contact that resulted from simulation and experiment of the artificial fingertip.

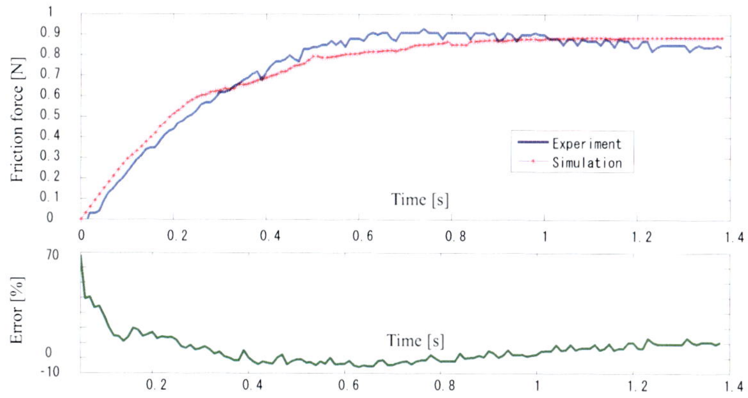

Fig. 4.20 Comparison of friction responses in simulation and experiment at $\theta = 15\,\text{deg}$, and $v = 6\,mm/s$

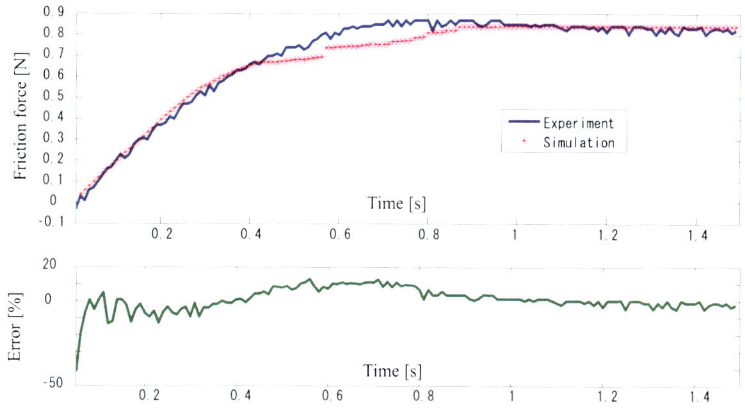

Fig. 4.21 Comparison of friction responses in simulation and experiment at $\theta = 45\,\text{deg}$, and $v = 5\,mm/s$

Fig. 4.20-4.21 show the agreement between two results of simulation and experiment, at different contact angle: 15 deg and 45 deg. The error is only significant at the initial time, then reduce to less than 10 % during the transition. The deflection at the simulation results was caused by artifact of numerical simulation. Both verified the fact that aforementioned in previous sections: there is a stick phase before the gross slide of the fingertip, during which the friction increases significantly (almost linearly). Plots in Fig. 4.22 indicate that when the inclination contact angle varies under the same contact depth and sliding velocity, the duration time of the stick phase also change correspondingly, in which smaller angle results larger duration of the stick phase. Similar conclusion can be observed in the experiment. It is due to the fact

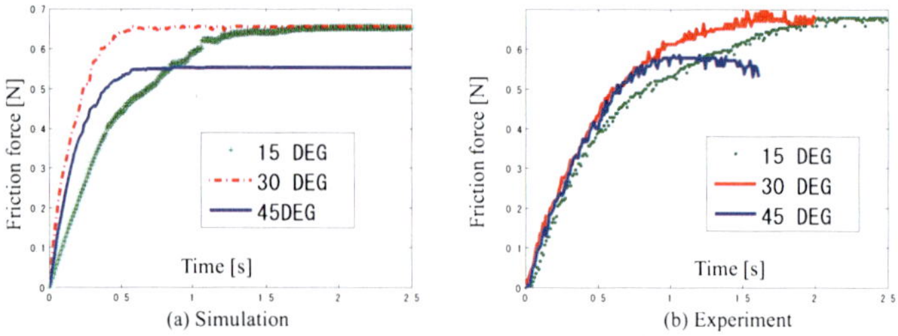

Fig. 4.22 Comparison of friction responses in simulation and experiment when the contact angle varies

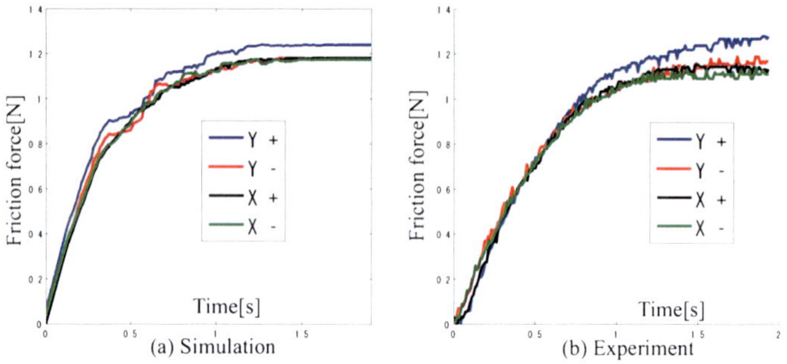

Fig. 4.23 Friction response when the fingertip slides over four directions as illustrated in Fig. 4.13

that at small contact angle the contact area is wider, resulting more time for the localized displacement to occur, also the friction is larger at the same contact depth and sliding speed.

We also investigate change at friction response when the fingertip moved in four different directions. We can observe in Fig. 4.23, in both simulation and experiment, that the friction does not vary significantly over four directions. Friction along $Y+$ direction is always slightly larger than those along remained directions. This conclusion is similar to what have been concluded in Section 4.6.2. This potentials the assumption that these slight differences are no enough for human to feel, and thus distinguishing directions of slide merely based on friction is difficult.

Localized Displacement Phenomenon

In order to verify this phenomenon, we mimicked method employed for human fingertip in Section 4.2.3 by marking a net of black dots on the outer skin of the artificial fingertip as illustrated in Fig. 4.16. The fingertip was then slid along X and Y direction.

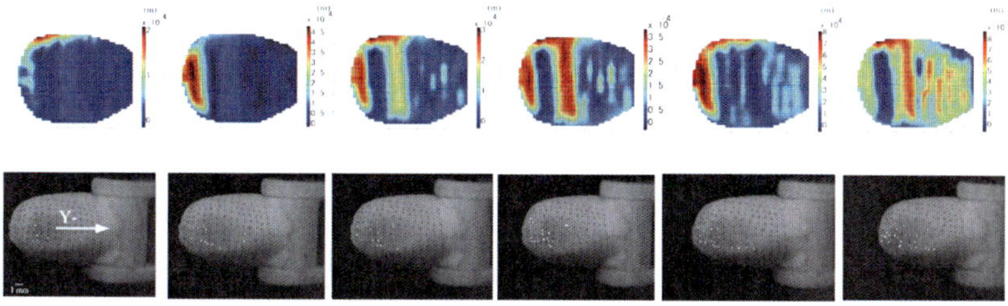

Fig. 4.24 Comparison in term of local displacement distributions over time when the fingertip slides along Y directions. Upper row is results from simulation.

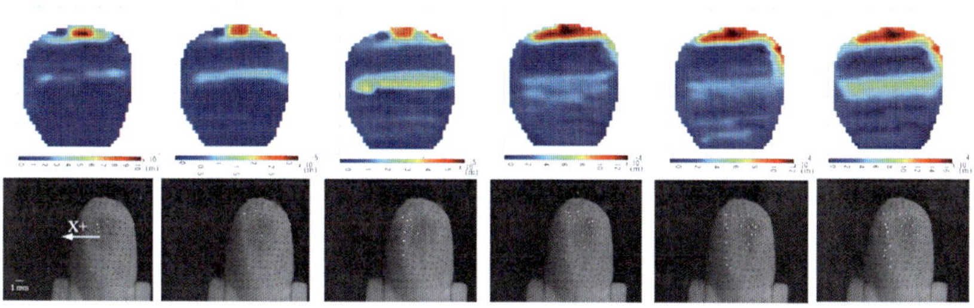

Fig. 4.25 Comparison in term of local displacement distributions over time when the fingertip slides along X directions. Upper row is results from simulation.

Fig. 4.24 and 4.25 illustrate partial movements of contacting areas in the contact pad during pre-slide phase obtained from both simulation and experiment along two distinguishing directions. While the movements of contacting points in experiment are indicated by white bars obtained from optical tracking method that is proposed in [42]; hot color in simulation represents larger displacement areas on the contact pad (note that the color scale is different over images.) We can observe that the propagation of slippage on the contact pad in simulation and experiment has obvious similarities in both cases. Boundary areas, where the normal force distributions are smaller according to Fig. 4.18, have slipped first before the propagation starts to erode the inner areas. We also can assess that the propagation of localized slippage varies significantly when the fingertip slides along different directions, which is similar to conclusion mentioned in Section 4.6.2. This, again, supports our recommendation that skin stretch caused by localized displacement is the main stimulus for mechanoreceptors underneath the skin of the human fingertip to feel tactile sensation regarding sliding action.

Consequently, we have verified the BBM in an artificial human-like fingertip with a fine experimental setup. Most of sliding mechanics' characteristics of the artificial fingertip are matched with those of simulated model, such as friction and the localized displacement phenomenon.

4.8 Case Study: Stable Grasping of an Object

In this section, we utilized the proposed model to simulate the process of lifting an object by two fingertips of human, taking into account the change of the localized displacement phenomenon as an input. Human, when attempts to lift an object with unknown weight, pays much less effort compared to robot, thanks to condense mechanoreceptors at fingertips that can detect vibration stimulated by the slippages so that human can slightly increase grip force to stably lift it, without harm it by hard gripping. It is said by neuroscience experiments that it is skin stretch caused localized slippage on the contact with object causes those stimulation to mechanoreceptors (see [49]). Therefore, by mimicking this process with the proposed model and a simple PD (Proportion-Derivative) control, further theoretical investigation can be expedited.

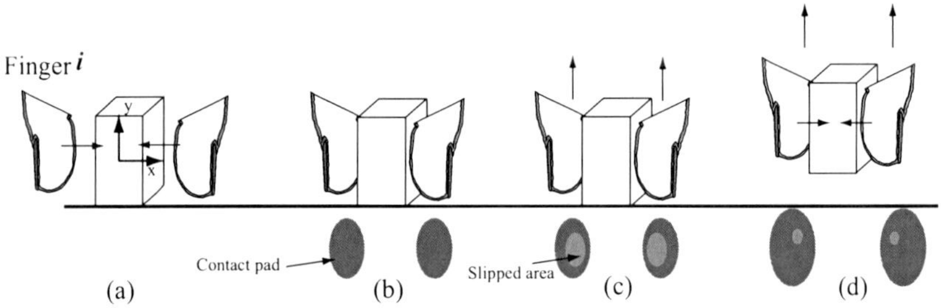

Fig. 4.26 Stable lifting process of human fingers

In this scenario, two identical human fingertips were approaching an object with an unknown weight. Let us assume that, for sake of simplicity, this object was a homogeneous rectangular cube. Human, with visionary assessing, could estimate where to grip the object so that two fingers were placed symmetrically on the cube's facets to eliminate unexpected rotation action (see Fig. 4.26(a)). Nonetheless, human cannot estimate the exact weight or fragileness of the object in order to generate suitable grip force. Therefore, he/she simply pinched the object by two fingertips (in this scenario) with a fair initial contact depth (Fig. 4.26(b)). Then, two fingertips moved up with a small velocity. During this phase, if slipped zone (indicated draftily by dark color in Fig. 4.26(c)) on the contact pad widened speedily, *i.e.* the object is likely to drop. Localized displacements on the contact pad expedited mechanoreceptors beneath the skin to send about-to-slip signal to human's brain. Human would increase the grip force enough to reduce the slipped area and enhance the friction in order to assure the object would not slip out of two fingers while lifting it (Fig. 4.26(d)).

4.8.1 Quantify the Slippage

To evaluate the level of slip on the contact pad, we introduced a quantitative value named as *slip indicator* λ, which is computed based on the ratio between the slipped area and the contact area:

$$\lambda = \frac{S_s}{S_c}. \tag{4.12}$$

Obviously, this indicator lies between 0, totally stick state, and 1, totally slip state. During the incipient slip, or pre-slide phase, this ratio varies between 0 to 1. Thus, we can choose this indicator as an feedback input of current stick/slip state of the contact area between a fingertip and the object's facet. However, it is not mandatory to force this value to critical value of 0 or 1 for stick or slip state correspondingly. A small slippage can be acceptable for safe lifting, meanwhile a quick rise of the indicator, even smaller than 1, can likely cause a slippage. In this simulation, we chose the largest value for stick phase is $\lambda_{stick} = 0.1$, and the smallest value for slip phase is $\lambda_{slip} = 0.5$. These are also two important values for the control of stable object lifting simulation.

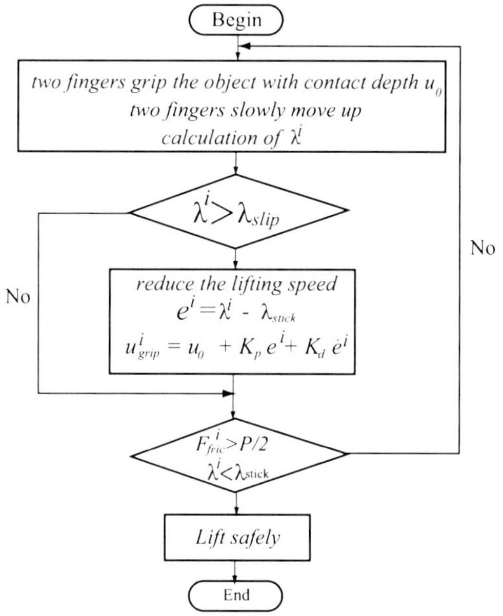

Fig. 4.27 Control flowchart

4.8.2 Control Method

Fig. 4.27 illustrates the flowchart of the simulation of the stable lifting scenario. First, two fingers grip the object with a pre-determined contact depth u_0, then start to move upward gradually. In this phase, the object does not move since the generated frictions on contacts with fingers are smaller than its gravity. According to the previous sections, two fingers start to deform and the localized displacements occurs on contact areas. If the total friction ($f_{fric} = \sum f^i_{fric}$, $i = 1, 2$) is not larger than the object's gravity P_{obj}, and the slip indicator of each finger λ^i is greater than λ_{slip}, say $\lambda^i > \lambda_{slip}$ ($i = 1, 2$), two fingers are likely to slip on the object's facets, thus a PD control is introduced to quickly increase the grip force to prevent slippages. Finally, when the total friction is larger than the object's gravity and the slip indicator falls to λ_{stick}, the object can be lifted safely with minimum grip force. In reality, since object's gravity is unknown, only the condition regarding slip indicator is necessary.

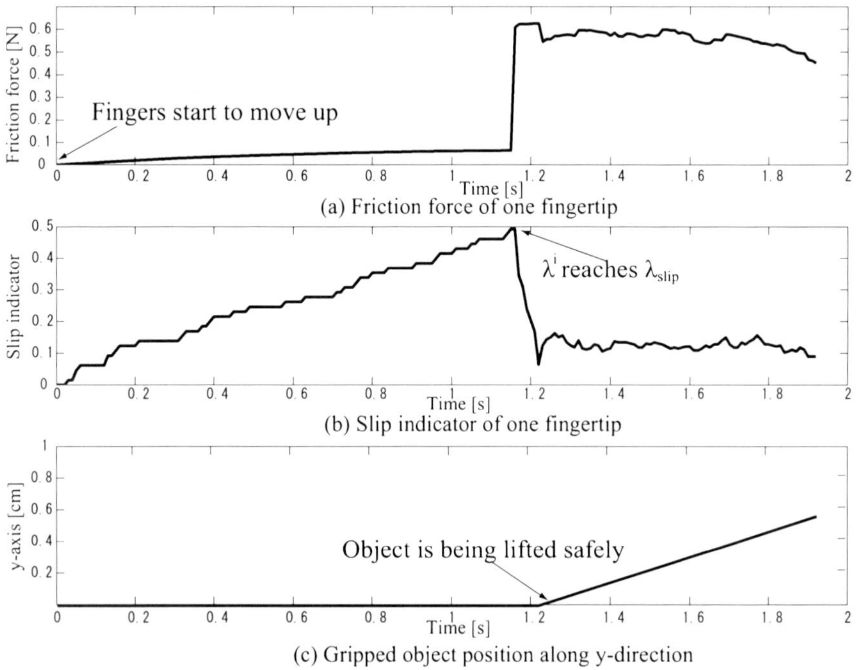

Fig. 4.28 Simulation result during the stable lifting

4.8.3 Simulation Result

In this simulation, the object's weight was 120 gr (1.2 N), friction coefficient between fingers and the object's facets is 0.6. For the PD control, $K_p = 3.0$

and $K_d = 1.0$. The initial contact depth u_0 was assigned bto0.8 mm. The speed of fingers was initially accelerated gradually. When the slip was about to occur, the speed was decelerated to reduce the possibility of sudden drop until the stable grip was assured by the PD control. Fig. 4.28 shows plots obtained during the stable lifting process of human fingertips. First, two fingertips start to move upward at the initial contact depth u_0, and the object position is still unchanged. During this phase, the friction generated at one fingertip is rising up gradually, in companion with remarkable increase of the localized displacement on the contact area indicated by the slip indicator λ (see Fig. 4.28(b)). When 50 % of the contact area had slipped, *i.e*, fingertips were likely to slip out of the object's facets, the stable gripping control was activated. Two fingertips increase the contact depth simultaneously, resulting a sudden jump of the friction as well as a quick decrease of the slip indicator, assuring that two fingers did not slide on the object's facets. The PD-control law ensured the slip indicator was around $\lambda_{stick} = 0.1$, *i.e* the object could be lifted up safely with the minimum gripping force. In this paradigm, we also added another condition that the total friction must be over the object's gravity 1.2 N, *i.e.* each finger's friction is over 0.6 N (see Fig. 4.28), before the object was lifted safely.

Consequently, the case study has shown the potential of our proposed model in studying stable grip/grasp mechanism of human soft-fingered hand. It also strengthens the selling-point that by observation of the localized displacement on the contact area, a stable grasp/manipulation can be assured with minimum grip force from human fingers. Further investigation can be developed based on our model, such as lifting irregular objects, stabilization of grasped object under disturbance, and so on.

4.9 Discussion

4.9.1 Significance of the LDP in Tactile Sensing

In this research, the proposed BBM suggested a way to model inhomogeneous soft fingertips in order to answer *how* and *when* the slippage happens during the stick phase through propagation of *localized displacements* on the contact area. This is helpful to intuitively image which phenomenon dominates sliding contact characteristics. Human never feels exact friction force acting on the contact area to judge tactile feelings. Instead, mechanoreceptors underneath the skin are stimulated by *skin deformation/stretch* through stress or vibration, then transferred to brainstem along axonal process [49], whereby tactile sensations are interpreted. This supports our idea of utilizing the LDP in detection of the incipient slip in tactile sensing system. It states that if a sensing system can detects the propagation of slippage on the contact area during the stick phase, in a *direct* or *indirect* way, the incipient slippage can be judged properly and *timely*. One of direct method is to use a high speed camera to track movement at each contact point, intuitively showing where the contact patch is sliding, where is not. Nonetheless, this method is not suitable for embedding in any system due to its size. Therefore, novel sensing system must be proposed for ease of fabrication and conforming to the LDP indirectly. The indirect method must take into account localized displacements into its transduced output so that robots/machines could assess the slippage is about to occur.

4.9.2 Beam Bundle Model

We have shown a process for application of the BBM in modeling various type of human-like fingertips, from actual one to artificial design, from homogeneous type to inhomogeneous one. Main characteristics of sliding human fingertip were successfully reflected in the simulated model, especially the localized displacements during pre-slide phase. The proposed method can act as a platform for studying sliding mechanics of sliding robotic fingertips, or inhomogeneous soft objects as a whole, in tactile perception, or creating models in virtual environment in haptic research. One can freely introduce various friction models into contact nodes such as LuGre, Dahl [39] models other than Coulomb friction one. Also, this model is convenient to calculate *friction moment* generated through sliding motion by synthesizing partial friction moments of contact nodes to a pre-determined axis. We are planning to create a bigger database of human fingertip, ranging from children to adult with different ages, to further investigate differences in sliding tactile mechanism, as well as enhance the proposed BBM. This model can be employed to model grasping/manipulation with friction of human or humanoid robots.

4.10 Concluding Remarks

We have theoretically investigated the sliding motion of a human fingertip model, focusing on the pre-slide phase, by utilizing a previously proposed beam bundle model and MR images. The simulation results help us to assess the change in force, and especially localized displacement phenomenon on the contact area during the pre-slide phase. The knowledge gained about this phenomenon will help us to understand the role of skin in the recognition of slippage, as well as to develop a sensing system to detect incipient slip of the human fingertip, something that is crucial in stable manipulation.

Part II

Tactile Sensing of Localized Slippage

In this part, we will show how the localized displacement phenomenon (LDP) and the proposed beam bundle model (BBM) can be applied in soft tactile sensing system to recognize the slippage. With beam bundle model, we can utilize to simulate and predict the output of transduced systems when a slippage occurs. Moreover, with the localized displacement phenomenon, ones can build their own sensing system based on this phenomenon. As a result, the BBM and the LDP could be exploited in analyzing sensor's output (Chapter 5), developing a novel sensor (Chapter 6), and enhancing performance of one off-the-shelf sensor with slip perception (Chapter 7). The content presented in this part strengthens the proposed theory mentioned in the previous part, helping the reader to imagine a complete loop from theory to application in the mechanics of sliding of soft objects.

5

Tactile Sensing via Micro Force/Moment Sensor

In previous chapters, we introduced a model of a sliding soft fingertip, and proposed the idea of LDP, which dominates the stick-to-slip phase of sliding motion, and is considered important in assessing slip detection in soft tactile systems. We can use this model to analyze the slip action of a soft tactile system, predict responses from sensors during sliding, and propose an efficient method of detecting slippage. In this chapter, we show our attempt to fabricate a tactile soft fingertip with an embedded MFMS. We employ a 2-D BBM to elaborate the slip action and corresponding responses from sensors, then utilize the LDP idea to propose a slip detection method. As mentioned in Chapter 1, tactile systems with slip perception have been developed over the past 20 years with various proposals regarding design and perception. There is a trade-off between the precision of physical quantity measurement and the speed of slip detection. A precise sensing system that can bring an exact exerted force/moment to act on a fingertip cannot detect slippage in a timely way since measurement is implemented mostly in the static state. Conversely, a dynamically responding tactile system that is suitable for the detection of slippage can only respond to physical phenomena, yet exact values. In this research, we aim to combine both above issues in one design with the assistance of the LDP.

We have focused on the development of a tactile sensing system and its application to tactile and texture recognition. The sensory core of this system consists of a MEMS technology-based 4-DOF MFMS chip, which can detect one component of force (F_z) and three components of moment (M_x, M_y, M_z) utilizing the piezoresistive effect in a single crystalline Si and on-chip Wheatstone bridge circuits. This sensing chip was developed by [52]. We implanted the micro sensor inside a hemispherical soft fingertip, which has been found theoretically to have simple and efficient object grasping control, to achieve a completed tactile sensing system. Our theoretical analysis focuses on assessment of the sliding motion of the fingertip, especially during the stick-to-slip phase. By performing both static and dynamic analysis, the responses of

A.-V. Ho and S. Hirai, *Mechanics of Localized Slippage in Tactile Sensing*, Springer Tracts in Advanced Robotics 99,
DOI: 10.1007/978-3-319-04123-0_5, © Springer International Publishing Switzerland 2014

the sensor during many phases of fingertip sliding can be predetermined, including the principle of incipient slip detection based on the LDP idea. The system was later carefully calibrated, and many experiments have been performed to show its potential in tactile and texture sensing, especially in incipient slip recognition. Consequently, the main contribution of this thesis is the proposal of a tactile sensing soft fingertip, and analysis to assess tactile perception during sliding motion.

5.1 Soft Fingertip with Micro Force/Moment Sensor

5.1.1 *Micro Force/Moment Sensor*

We utilized a 4-DOF MFMS chip that can independently detect one component of force (F_z) and three components of moment (M_x, M_y, M_z). The inset picture in Fig. 5.1(a) shows the fabricated MFMS chip measuring 2 mm x 2 mm x 0.5 mm. This MFMS has four crossed beams (Fig. 5.1(a)) with piezoresistors diffused at suitable places on the beams. The application of an external force to the chip deforms the four beams and changes the resistances of the piezoresistors, leading to changes in the output voltages of the corresponding measurement circuits. Four piezoresistors were placed on the surface of each beam, which can be categorized as those detecting longitudinal (normal) and shear stresses (Fig. 5.1(a)). The detailed principles of operation can be found in [52]-[53]. In this study, we configured the sensor construction so that it can output three components of force and moment, F_z, M_x, and M_y, by arranging measurement circuits on the outer frame of the sensing chip. Figure 5.1(b) shows three Wheatstone bridges, which were designed earlier, on this MFMS chip to convert change in resistance into an output voltage measuring the force F_z, as well as M_x, M_y. Integrating the Wheatstone bridges on the chip means there is no need for external transducer circuits, which keeps the implementation simple. Figure 5.1(b) (upper) shows a schematic of a half-bridge for measuring F_z, with two identical constant reference resistors R_{ref}, and four piezoresistors $R_{F_{z1}}$through $R_{F_{z4}}$. These two reference resistors were placed in a non-stress area and chosen so that the bridge is balanced in a non-stress state. When the resistances of the piezoresistors $R_{F_{zi}}$ change due to stress, the output voltage can be expressed as

$$V_{Fz} = \frac{r}{(1+r)^2}\left(\frac{\Delta R_{F_{z1}} + \Delta R_{F_{z4}}}{R_{F_{z1}} + R_{F_{z4}}} - \frac{\Delta R_{F_{z2}} + \Delta R_{F_{z3}}}{R_{F_{z2}} + R_{F_{z3}}}\right) V_{in}, \tag{5.1}$$

where

$$r = \frac{R_{F_{z1}} + R_{F_{z4}}}{R_{F_{z2}} + R_{F_{z3}}}. \tag{5.2}$$

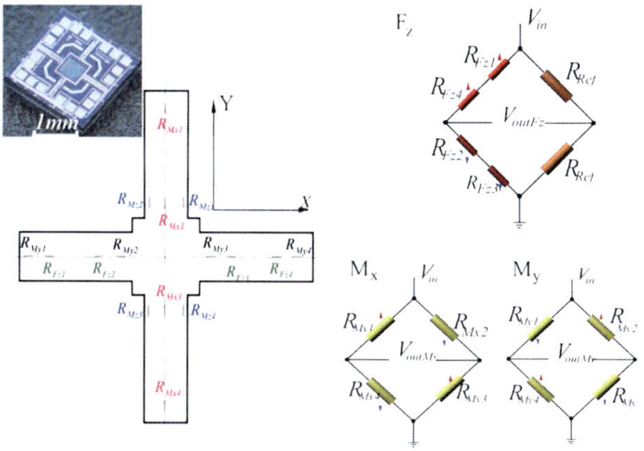

(a) Structure of the MFMS (b) Wheatstone bridges

Fig. 5.1 Four crossed beams of the MFMS chip with 16 piezoresistors. Each group of piezoresistors, corresponding to each component of force/moment, was wired to form a Wheatstone bridge to output the electrical signals based on changes in resistance of the piezoresistors. The inset picture shows an image of a fabricated MFMS chip.

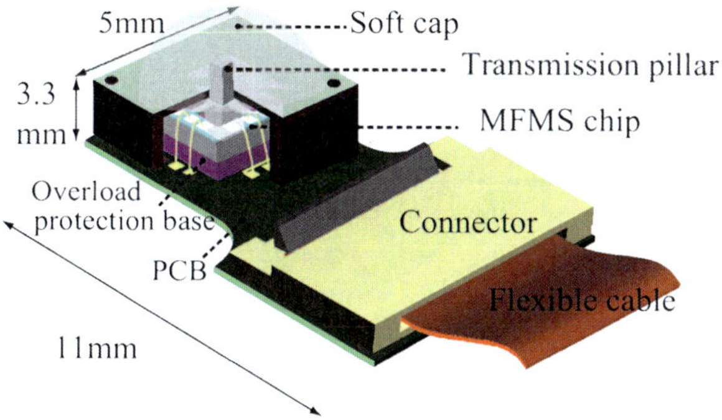

Fig. 5.2 Complete packaged micro force/moment sensor

Piezoresistors R_{Fzi} $(i = 1-4)$ are designed to be identical, thus $r = 1$. When a vertical force F_z is applied to the center of the sensing chip, the longitudinal stresses in all the piezoresistors of the F_z bridge can be written as:

$$\sigma_{R_{F_{z1}}} = \sigma_{R_{F_{z4}}} = -\sigma_{R_{F_{z2}}} = -\sigma_{R_{F_{z3}}}, \tag{5.3}$$

where $\sigma_{R_{F_{zi}}}$ is the longitudinal stress at piezoresistors $R_{F_{zi}}$ $(i = 1-4)$. Therefore, the following relationship is satisfied:

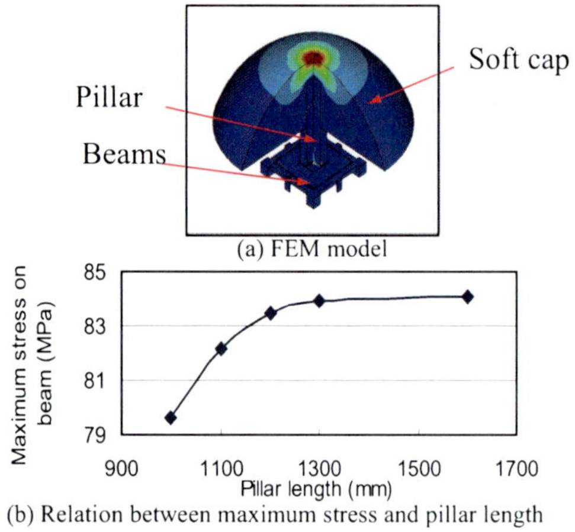

Fig. 5.3 Simulation model (top) and relationship between length of the pillar and maximum stress (bottom)

$$\Delta R_{F_{z1}} = \Delta R_{F_{z4}} = -\Delta R_{F_{z2}} = -\Delta R_{F_{z3}}. \tag{5.4}$$

Finally, the output voltage of the F_z bridge can be expressed as:

$$V_{outFz} = \frac{1}{2}\frac{\Delta R_{F_{z1}}}{R_{F_{z1}}} V_{in}. \tag{5.5}$$

In addition, due to the symmetry of the arrangement of the piezoresistors and the structure of the sensing chip, the resistance change in each beam will satisfy the equation $\Delta R_{M_{y1}} = -\Delta R_{M_{y2}} = -\Delta R_{M_{y3}} = \Delta R_{M_{y4}}$. Therefore the M_y-bridge is still balanced. Similar results were observed with the M_x-bridge, thus eliminating crosstalk between the bridges. The measurement principles were similar for the M_x, and M_y Wheatstone bridges (Fig. 5.1(b)).

After constructing the cored MFMS chip, a transmission pillar, made from silicon material and designed to receive an external force from a silicon rubber tip and transfer it to the crossed beams, was placed onto the center of the chip. An FEA simulation in Fig. 5.3 shows the relationship between the pillar and the maximum stress on a beam when a force $F_z = 0.4$ N is applied to the top of the cap. We observed that at a length of 1.6 mm, the stress value was almost saturated at 84 MPa. As a result, the size of the pillar was set at with 400 μm x 400 μm in cross section and 1.6 mm in length. We also analyzed this situation without a pillar to show the advantages of using the pillar to transfer external force to the beams, with the maximum stress on the beams being 250 MPa. Therefore, by using a pillar 1.6 mm in height and a soft tip, the amount of stress acting on the beams was four times lower without loss of stress.

Next, the entire MFMS was placed on a base made of Pyrex glass to protect it from overloading. Gold wire bonding was used to connect the chip to a

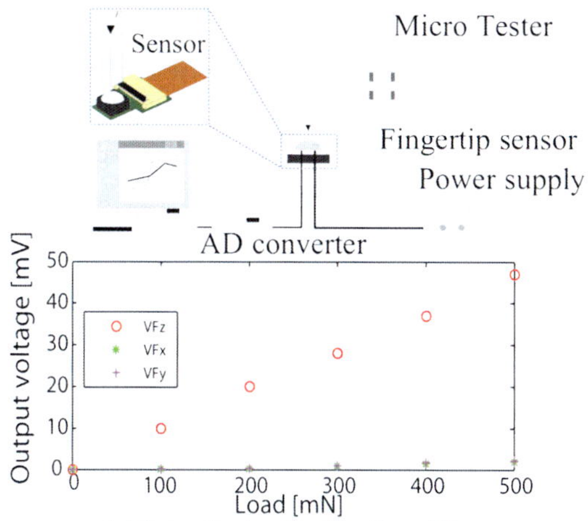

(a) Setup (top) and calibration of Fz

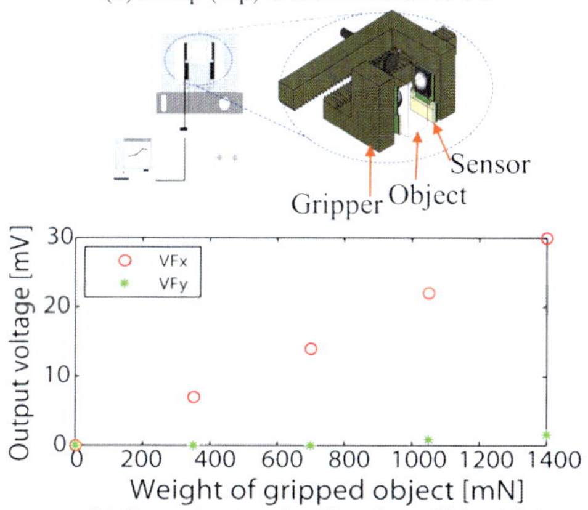

(b) Setup (top) and calibration of Mx (My)

Fig. 5.4 Calibration setups for the sensor

printed circuit board (PCB) using a bonding machine, and both the chip and the protective base were enclosed in a ceramic case. A flexible ribbon cable was used to direct the input and output voltages (Fig. 5.2). The overall dimensions of the packaged sensor were 11 mm in length, 5 mm in width, and 3.3 mm in thickness. The sensor was calibrated using an INSTRON micro tester. The experimental setup for the calibration of F_z is shown in Fig 5.4(a). A calibration force F was applied to the top of the cap, and the outputs were recorded. At a voltage of 3 V, the sensitivity to the F_z component (SF_z) was 0.085 mV/mN,. M_x and M_y were calibrated by mounting two micro sensors onto two opposing fingers of a robot hand, and conducting a gripping operation, as shown schematically in Fig. 5.4(b). The gripping force (F_z) was adjusted so that sliding of the gripped object did not occur. Aluminum blocks of known weights were used as the objects. The length of the moment

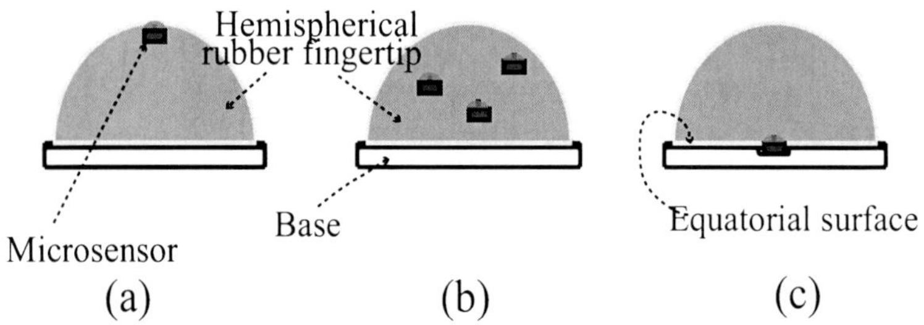

Fig. 5.5 Possibilities for embedding a sensor inside a hemispherical soft fingertip: (a) Near the outer skin of the fingertip. (b) Within the fingertip. (c) At the center of the equatorial cross-sectional area.

arm was predetermined to be the pillar length. The relationship between the weight of the gripped object and the output voltage, VF_x, was derived as shown in Fig. 5.4(b). The sensitivity of F_x and F_y (SF_x) was 0.039 mV/mN. The effective range of the sensing chip was based on the linear range of piezoresistive effectiveness. The upper limit of stress that the sensor was able to support was 500 MPa, corresponding to a maximum normal force of 0.8 N and a maximum tangential moment of 2 Nmm. The crosstalk was less than 10 %, as observed in the results of calibration experiments. Repeatability, calculated by computing the mean error when applying five different loads, was about 4.2 %. The sensor exhibited a low hysteresis of about 8 %, and low time drift of 4.5 % for one hour.

5.1.2 Encapsulation

Soft membranes have been used to cover sensors, artificially simulated human skin [21]-[82], and human fingertip-like shapes [83]-[54]. Previous research on hemispherical soft fingertips showed that during dynamic simulation of object manipulation using only two 1-DOF soft fingertips and the local minimum elastic energy (LMEE) theory, object posture control was achieved easily and stably, while at the same time keeping control simple [10]. We therefore decided to use hemispherical soft fingertips as a soft compliant encapsulation material for the tactile sensing system. These fingertips were made of polyurethane rubber fabricated using a casting process and were 20 mm in diameter, similar to the tip of an adult thumb.

That left the question of where to place the micro sensor inside the soft fingertip. Figure 5.5 shows some options for embedding the sensor inside the fingertip. If the sensor is placed near the outermost border of the fingertip (Fig. 5.5(a)), the actual forces acting on the fingertip would be the same as on the sensor; therefore the sensor would be highly sensitive to the states of contact between the fingertip and the outside world. However, as the maximum normal load on the pillar is less than 0.8 N, placing the sensor on the outermost border may lead to its breaking if the fingertips handle a heavy

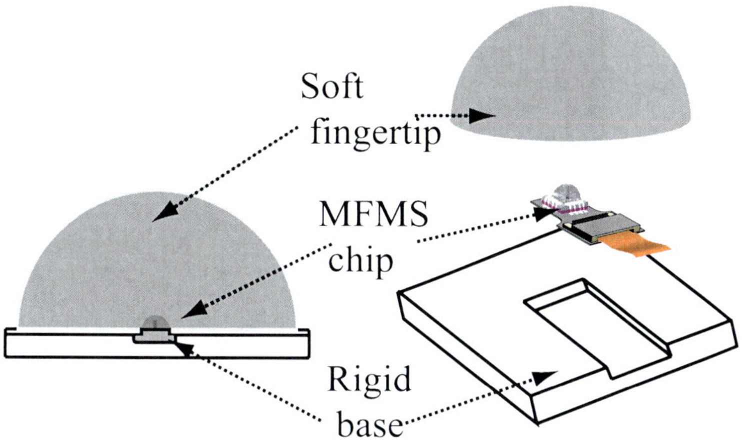

Fig. 5.6 Packaging of the entire tactile sensing soft fingertip

object or are overloaded. Furthermore, as the packaged sensor has dimensions of 11 mm in length, 5 mm in width, and 3.3 mm in height, sufficient space is required inside the soft fingertip to accept the sensor. This may cause complications in fabrication and in terms of signal wires, as well as as affecting the proper deformation of the soft fingertip. If the micro sensors are arranged as illustrated in Fig. 5.5(b), the number of drawbacks would be similar to those in Fig. 5.5(a). In addition, such an arbitrary placement of the sensor would not efficiently reflect the state of contact of the soft fingertip [54]. In Fig. 5.5(c), the sensor is placed at the center of the equatorial cross-sectional area of the hemispherical fingertip. In this configuration, the volume of the sensor is completely unrelated to the volume of the fingertip. In addition, there are fewer complications in fabrication. Since this arrangement reflects the natural deformation of the soft fingertip without affecting sensor size, it is considered appropriate. The signals of force and moment from the sensor correspond to those acting at the center of the equatorial surface of the fingertip, but these are not the actual external force/moment acting on the outer surface of the fingertip. In conclusion, we chose the configuration in Fig. 5.5(c) to form a tactile sensing soft fingertip.

To form a complete system, the following process was implemented. After fabrication of the micro sensor, it was encapsulated. First, a square base (20 mm × 20 mm × 5 mm) made of acrylic resin ACBTA (Misumi, Japan) was made; this base contained a gutter, 10 mm × 7 mm × 3 mm, on which the micro sensor was placed so that the position of the pillar was coincident with the central point of the upper surface of the base. Subsequently, KE-12 polyurethane (Exseal, Japan) was poured into a hemispherical mold (radius r = 10 mm) and allowed to cure for 5 hours at room temperature. Finally, this polyurethane rubber hemispherical fingertip was attached to the surface of the acrylic base to which the sensor had been attached (Fig. 5.6).

5.2 Analysis of Micro Force/Moment Sensor Performance

5.2.1 Static FE Analysis

Modeling: To determine the stress distribution of a soft fingertip when it slides, and to investigate the response of the sensor, a model of a hemisphere was constructed using FEA, consisting of deformable material, a rigid surface, and corresponding loads. Although FEA has been used previously to simulate deformation of a soft hemisphere [7], that deformation was caused only by pushing vertically on a rigid surface for sake of modeling and simulation.

The material we used to make the soft fingertip was polyurethane rubber, an elastomer material with elastic properties. Therefore, for similarity, we used a model of a hemisphere formed of hyperelastic material similar to an elastomer material. The Mooney-Rivlin model is the mathematical model for hyperelastic material used for simulation, calculated from the stress-strain characteristics of polyurethane. Other requirements of the model material, such as incompressibility, homogeneity and isotropy, were also included in the simulation. The meshed FEA model of a hemisphere with a diameter of 20 mm is composed of 15209 nodes. The element type of this model is SOLID 186, a 3-DOF 20-node structural solid. This element has plasticity, hyperelasticity and the ability to withstand large strains, and each node has 3-DOF. The FEA model of the surface is much simpler, containing only rigid elements with 1-DOF (translation). Determination of the deformation of the fingertip requires determination of the relative movement between the hemisphere and the surface. To do so, we set **A** as the equatorial surface of a hemisphere that can translate vertically only along the z-axis and is constrained from other translations or rotations; and **B** as a rigid contact surface that can only translate horizontally along the x-axis (Fig. 5.7(a)). The surface **A** forms a tilting angle α with the surface **B**. Sliding occurs with a predetermined coefficient of friction of 0.5. The FEA model employed here is a nonlinear model due to the nonlinear material, the contact problem, and the large deformation in the model. The coordinates of measurement of the sensor O-xyz are attached in a fixed position to the center of the equatorial surface of the soft fingertip. This model was examined under situations in which the fingertip is moved along two opposite directions, the $(+)$*direction* and $(-)$*direction*, as illustrated in Fig.5.7(b). Simulation results are calculated using these coordinates.

Deformation of the soft fingertip was assessed in the $(+)$ direction, as shown in Fig. 5.7(b). This simulation also shows stress distribution on the contact surface, as illustrated in Fig. 5.8(a). On this surface, the shape of the

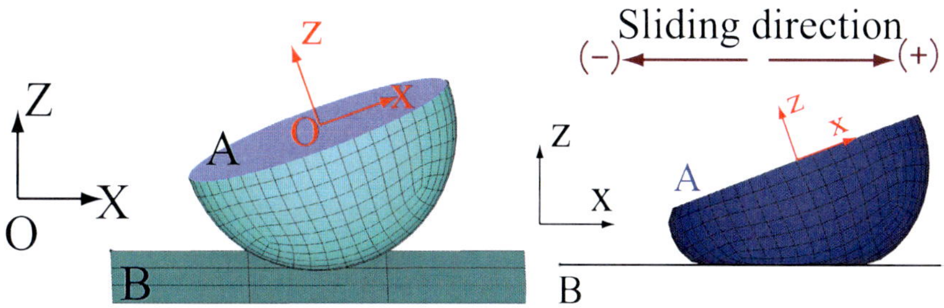

(a) Meshed model.

(b) Deformed shape.

Fig. 5.7 FE analysis model for simulating the deformation of a soft fingertip

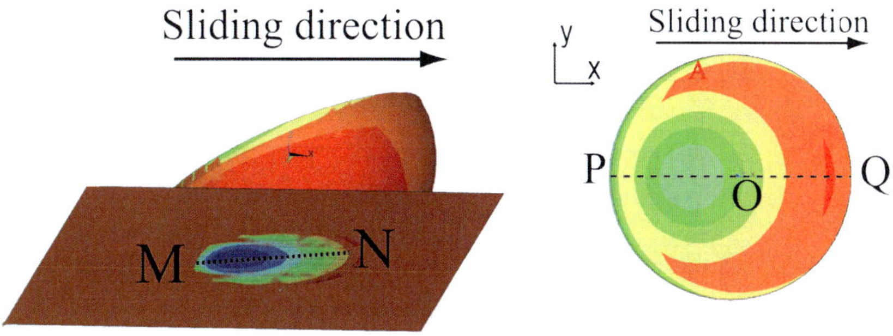

(a) On the contact surface.

(b) On the equatorial surface **A**.

Fig. 5.8 Stress distributions on two surfaces

entire contact region was similar to an ellipse, with normal and shear stress distributions forming many elliptical rings with their centers shifted along the direction opposite to sliding. Figure 5.8(b) showed the corresponding stress distribution on the equatorial surface **A**, with the central point O, where the sensor is located. We observed a similarity for these two distributions. More detailed information can be obtained by determining the normal forces distributed on the path PQ of the equatorial surface **A** (Fig. 5.8(a)), and on the path MN (Fig. 5.8(b)) of the contact surface. The micro sensor is located at the center O of path PQ, with the force acting on this point equal to that acting on the sensor. Figure 5.9(a) shows the normal force distributions on path PQ and MN when the soft fingertip (FT is pushed 2 mm in contact depth, and when the fingertip is simultaneously pushed 2 mm vertically and moved horizontally with various displacements. The length of PQ is 20 mm, while the length of MN depends on the contact depth and sliding displacement of the fingertip on the surface. The shapes of the distributions on PQ and MN are coincident. When the FT is moved gradually (note that the contact surface still sticks to the plane), these distributions shift along the direction

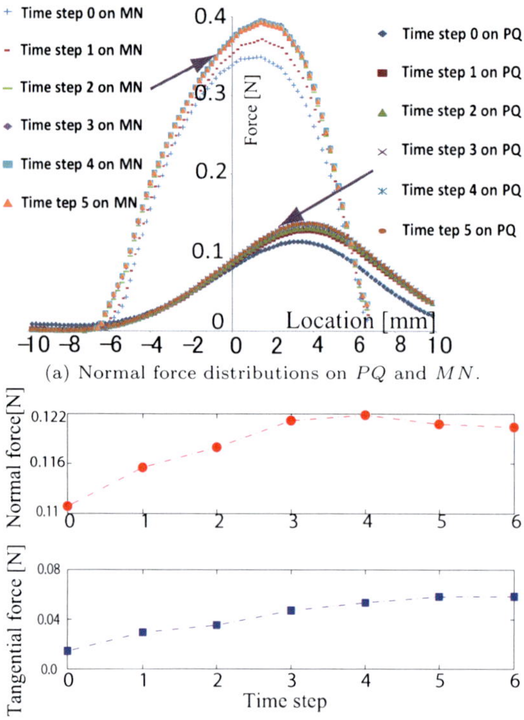

(a) Normal force distributions on PQ and MN.

(b) Normal and tangential forces at point O of path PQ over time steps.

Fig. 5.9 Simulation results when the fingertip slides along the (+) direction

opposite to sliding, and the value of the normal forces at the center point of PQ and MN increases. These augmentations cease simultaneously, decreasing the values of the normal forces. In contrast, by determining the shear force acting on the center point of the PQ path, we found that this value increased when the FT was moved, becoming maximal at slippage. Moreover, the times at which the lowest normal force and the highest shear force occur differed (Fig. 5.9(b)). Interestingly, this phenomenon changes completely when the FT moves along the (−) direction. As shown in Fig. 5.10(a), distributions reallocate opposite to those observed in the (+) direction. The normal force at the center of path PQ decreases, reaches its minimum and increases; while the tangential force keeps increasing and becomes stable (Fig. 5.10(b)). Delays in time, similar to those observed in the (+) direction, were also seen. This phenomenon is easier to observe if the tilting angle $\alpha = 0\,\text{deg}$ (Fig. 5.11). In that case, the distributions of stress are symmetric, and the maximum values are at the center points. When the FT moves in either direction, the points of maximum normal force are no longer at the center, but are shifted to other points on the two paths, (PQ and MN), causing normal forces to decrease while shear forces increase. Thus, if the sensor is located at the center O of

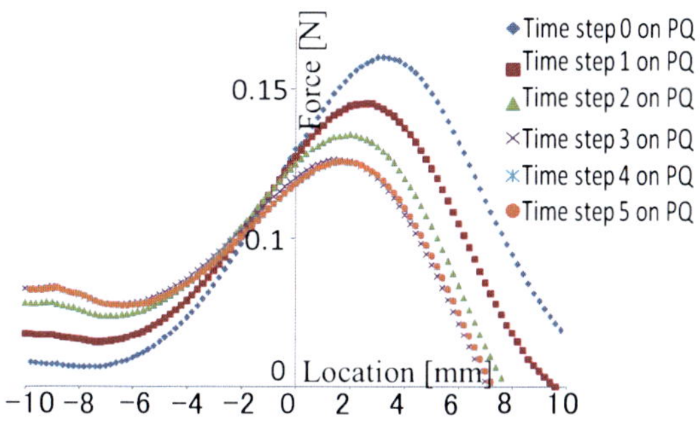

(a) Normal force distributions on PQ.

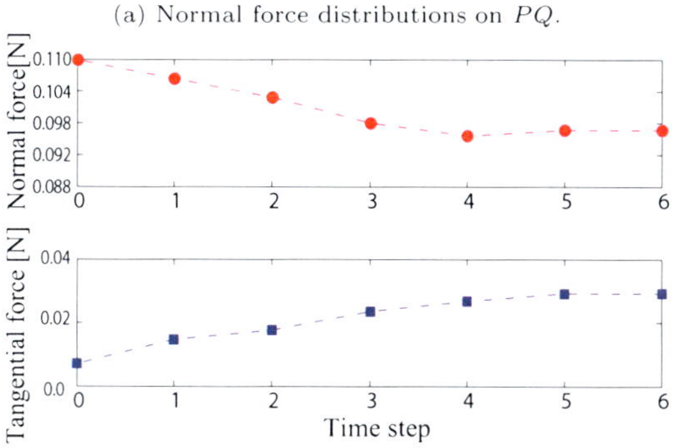

(b) Normal and tangential forces at point O of path PQ over time steps.

Fig. 5.10 Simulation results when the fingertip slides along the (−) direction

path PQ, its normal force responses (F_z), and moment M_x, M_y will change when the state of contact between the fingertip and the surface changes. Due to the static simulation, however, only the latter results were obtained, and not only during the sliding process. Therefore, the results in Fig. 5.9(b) or Fig. 5.10(b) cannot determine when and exactly how the moment of slip of the fingertip occurs or detailed changes in normal and shear forces.

Consequently, FEA of a sliding soft fingertip can determine stress distributions on the contact surface, as well as at the sensor location. For each specific state of contact of the fingertip, the stress differs, depending on the location of the sensor. Before the onset of gross sliding of the fingertip, normal forces increase or decrease, while tangential forces increase. We can even distinguish the direction of slide based on the trend of change of normal force. This suggests a method to detect incipient slip, which is important in stable grasping. However, this simulation cannot yield a *dynamic* picture at how the slippage happens over time, thus for those who expect better description of slip occurrence can use our model as a countermeasure.

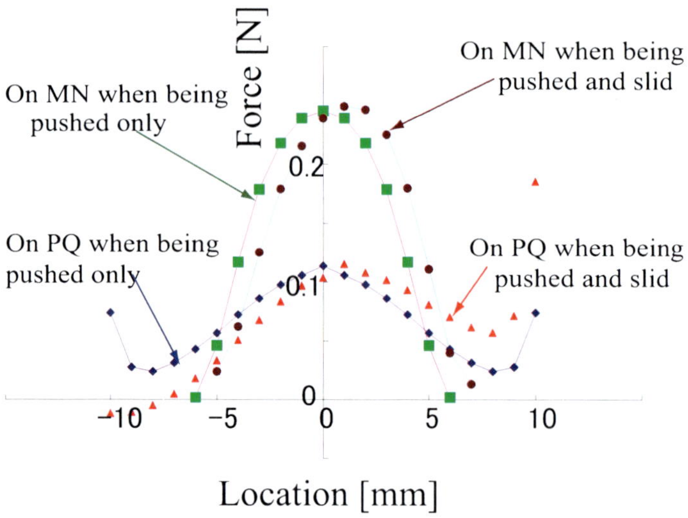

Fig. 5.11 Normal force distributions MN and PQ paths when the finger starts to slide at $\alpha = 0\,\text{deg}$

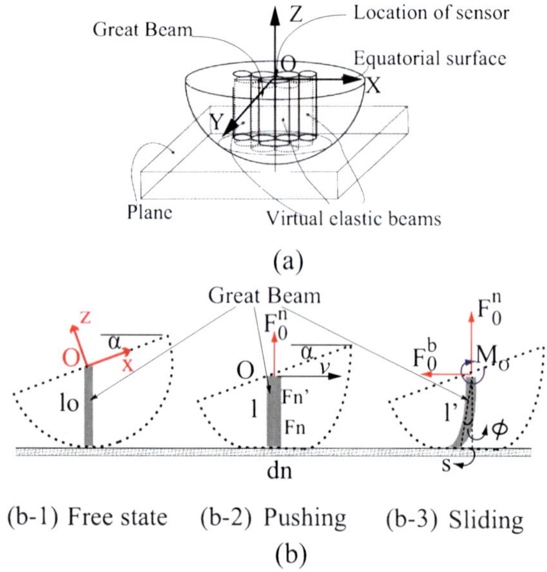

Fig. 5.12 (a) Model of a soft fingertip with a finite number of elastic beams. (b) Corresponding deformed states of the Great Beam whose free end is the location of the sensor.

5.2.2 *Dynamic Analysis Using BBM*

Modeling

In this section, we utilize our proposed BBM reported in previous chapters with some modifications to simulate a sliding fingertip, to calculate the corresponding responses of the embedded sensor from which a method to detect slippage can be proposed. We also see that the LDP plays an important role in assessing the responses of the embedded sensor during sliding motion of the soft fingertip.

Assuming that a fingertip is composed of a finite number of virtual flexible beams with their fixed ends on the equatorial surface (O-XY) and their free ends on the outer surface (Fig. 5.12(a)), deformation of this fingertip would be represented by the deformation of all beams, as well as by elastic forces and elastic energy [according to BBM, see previous chapters]. Therefore, the sensor is assumed to be placed at the fixed end O of the beam at coordinates $(0, 0, 0)$; hereafter, this beam will be called the *Great Beam*. The fixed end is fixed relative to the rigid base so that it can move when the rigid base moves. We could therefore approximate the forces and moments acting on the sensor as those acting at the fixed end of the Great Beam. When the Great Beam is compressed, stretched, or bent by an external load, we can determine the forces and moments at the fixed end, *i.e.* the sensor, appropriately (Fig. 5.12(b)). For the sake of simplicity, we investigated the interaction of the Great Beam with neighboring beams in the great cross-sectional area of the soft fingertip. The fingertip makes a titling angle α when in contact with the surface.

In our model, we assumed that those beams interacted with one another via*linkage* springs (with stiffness k) and dampers (with damping coefficient d) on the contact surface; and that the free ends of the beams not in contact with the surface were irrelevant to the sliding motion. Analysis of the force/moment acting on the model is illustrated in Fig. 5.13. If there are $2N + 1$ contacting beams, numbered $i \in [-N, N]$ along the direction of slide, The number N was based on the depth of contact d_n and beam parameters, such as natural length l_i and the cross-sectional radius r_i. If $\mathbf{u} = [u_{-N}, .., u_i, .., u_N]^T$ is a collective vector of the generalized displacement of the free ends of all beams, then F_i^n, F_i^{fr}, and F_i^b are the normal, frictional, and bending forces, respectively, at the free end of the i-th beam. In addition, there are interaction forces F_{i-1}^{int} and F_i^{int} between the free end of the i-th beam and the two neighboring free ends. When the fingertip is pushed to a depth d_n (Fig. 5.12(b-2)), the normal force acting at the free end of an arbitrary beam with coordinates (x,y) can be expressed as:

$$F_i^n = E\frac{s_i}{l_i}\left(\sqrt{r^2 - (x^2 + y^2)} - \frac{r - d_n - x\sin\alpha}{\cos\alpha}\right), \qquad (5.6)$$

where E is Young's modulus. As a result, the normal force acting on the free end of the Great Beam, *i.e.* $x = 0$ and $y = 0$, is as follows:

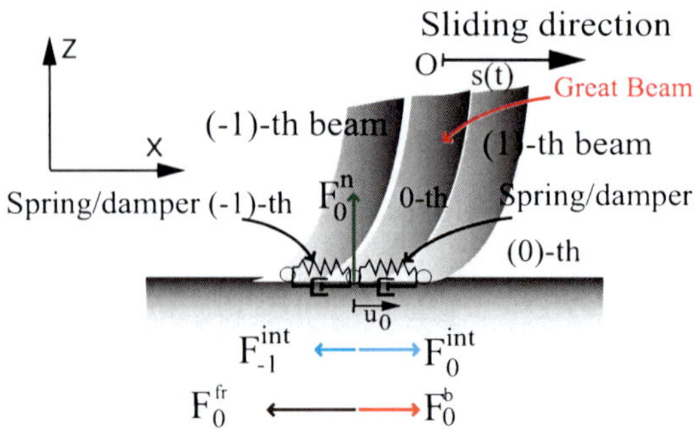

Fig. 5.13 Analysis of force/moment acting on the Great Beam and its neighboring ones

$$F_0^n = \frac{Es_0}{l_0} d_n = \frac{Es_0}{l_0}(l_0 - l). \tag{5.7}$$

The bending force is based on the bending strain s of beams when the FT moves but the contact surface still sticks to the surface, as illustrated in 5.12(b-3). The value of the bending force is obtained as:

$$F_b^i = k_i^b s = \frac{3EI_i}{(l_i)^3} s = \frac{3E\pi(r_i)^4}{(l_i)^3} s, \tag{5.8}$$

with k_i^b being the bending stiffness, and I_i being the moment of inertial of a beam [28]. The two interacting forces, F_{i-1}^{int}, and F_i^{int}, were caused by deformations of the $(i-1)$-th and i-th linkage spring/damper elements when the free end of the i-th beam starts moving; and can be described as:

$$F_i^{int} = k(u_{i+1} - u_i) + d(\dot{u}_{i+1} - \dot{u}_i), \tag{5.9}$$

and

$$F_{i-1}^{int} = k(u_i - u_{i-1}) + d(\dot{u}_i - \dot{u}_{i-1}). \tag{5.10}$$

For the Great Beam, i equals zero. Thus, if l_0, l, and l' are the lengths of the Great Beam in the natural, pushing, and sliding states, respectively (Fig. 5.12(b)), and F_O and M_O are the force and moment, respectively, acting at the fixed end O of the Great Beam along the z- and x-axes of the sensor's measurement coordinates O-xyz (Fig. 5.12(b-1)), then the reaction force acting on the fixed end of the Great Beam, including the normal compressive and bending forces, can be calculated relative to the sensor's coordinates as:

$$F_O = \begin{bmatrix} \cos\alpha \ p\sin\alpha \end{bmatrix} \begin{bmatrix} F_{i=0}^n \\ F_{i=0}^b \end{bmatrix} = \begin{cases} \cos\alpha F_0^n + \sin\alpha F_0^b & \text{if } p = 1 \text{ direction } (+) \\ \cos\alpha F_0^n - \sin\alpha F_0^b & \text{if } p = -1 \text{ direction } (-) \end{cases} \tag{5.11}$$

When O starts to move with the predetermined velocity v, the free end still sticks to the ground, and the beam is bent, causing a change in the length of the beam (Fig. 5.12(b-3)). As a result, the normal force is calculated as:

$$F_0^n = \frac{Es_0}{l_0}(l_0 - l') = \frac{Es_0}{l_0}\left(l_0 - \frac{l}{cos\phi}\right), \tag{5.12}$$

while ϕ is calculated as:

$$\phi = \arctan\frac{s(t) - u_0(t)}{l}, \tag{5.13}$$

where $s(t)$ is the distance moved by point O and u_0 is the generalized displacement of the contact point of the Great Beam at time t. As a result, the normal force changes during the sliding motion of the fingertip, due to deformation. This is slightly different to the models mentioned in the previous chapters, in which normal force distribution is assumed to be consistent. The moment acting on the fixed end of the Great Beam consists of two components around the x and y axes, M_{Ox}, and M_{Oy}, respectively. $M_{Ox} = 0$ due to the symmetry around the x-axis when the fingertip slides along this axis, whereas M_{Oy} is more complicated. The latter includes moments caused by frictional forces and the unsymmetrical distribution of normal forces acting on the free ends of beams. As a result, the moment acting on the fixed end of the Great Beam is calculated as:

$$M_O = M_{Oy} = \sum_{i=-N}^{N} F_i^{fr} l + \sum_{j=-N}^{0} F_j^n x_j - \sum_{k=0}^{N} F_k^n x_k, \tag{5.14}$$

or

$$M_O = M_{Oy} = \sum_{i=-N}^{N} F_i^{fr} l - \text{sign}(i) F_i^n x_i \tag{5.15}$$

In the above equations, the interactive force couple F_i^{int} and F_{i-1}^{int} are internal ones that vanish in summation.

As we have proposed a method to derive the frictional force acting on the free ends of the beam during the stick-slip phase in the previous chapters, if $\mathbf{B} = [k_{-N}^b, .., k_i^b, .., k_N^b]^T$ is the vector of bending stiffness of the beams, then the motion equations for all the beams, *i.e.* a soft fingertip, during sliding motion can be summarized as:

$$-\mathbf{Ku} - \mathbf{D\dot{u}} - \mathbf{F^{fr}} + \mathbf{B}s(t) = m\mathbf{\ddot{u}}, \tag{5.16}$$

where $\mathbf{F_{fr}} = [F_{-N}^{fr}, .., F_i^{fr}, .., F_N^{fr}]^T$, and:

$$\mathbf{K} = \begin{pmatrix} b_{-N} + k & -k & & \\ -k & b_{-N+1} + 2k & -k & \\ & \vdots & \vdots & \vdots \\ & & -k & b_N + k \end{pmatrix},$$

Table 5.1 Parameter Values Used in Simulation

Parameter	Value	Unit
r	10	mm
d_n	2	mm
d	1	mm
2N+1	13	beams
k	650	N/m
d	1.6	Ns/m
m	0.002	kg
μ	0.5	
α	10	deg

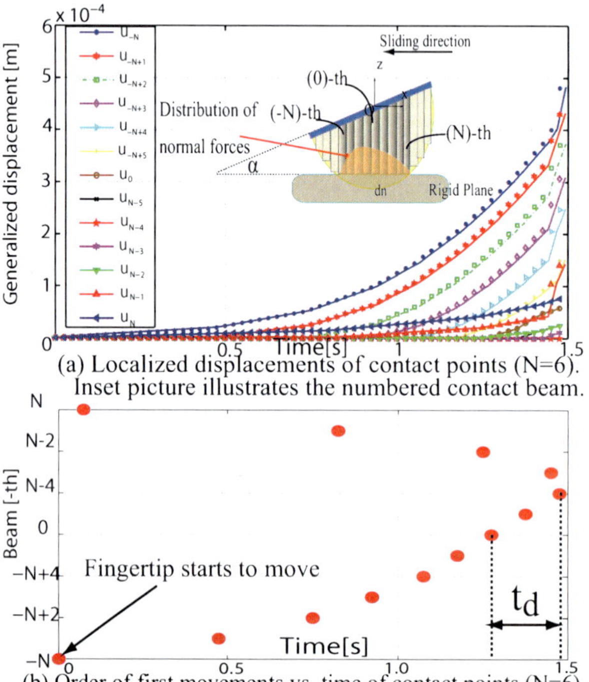

(a) Localized displacements of contact points (N=6). Inset picture illustrates the numbered contact beam.

(b) Order of first movements vs. time of contact points (N=6)

Fig. 5.14 Movements of contact points during the stick-slip phase along the (−) direction. The number of contact beams N is set to 6.

$$\mathbf{D} = d \begin{pmatrix} 1 & -1 & & & \\ -1 & 2 & -1 & & \\ & \vdots & \vdots & \vdots & \\ & & & -1 & 1 \end{pmatrix}$$

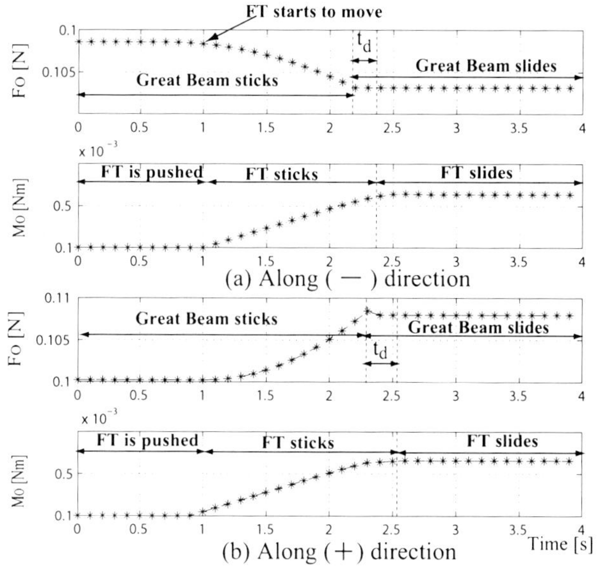

Fig. 5.15 Plots of force F_O, and moment M_O on the fixed end of the Great Beam, *i.e.* location of the sensor, during the stick-slip phase of the fingertip (FT) in two directions: (a) Direction (+). (b) Direction (−).

are referred to as the contact stiffness and contact damping matrices, respectively. We used a numerical method to assess movements. During the simulation, we calculated the frictional force acting on the free end of each beam and used it to determine whether the free end sticks or slips. The overt slip of a soft fingertip on the contact surface will occur when all of its free ends slide.

Simulation Results

A simulation was conducted under the same conditions as in Section 5.2.1. The other parameters are summarized in Table 5.1. First, the FT moves along direction (−) at tilting angle $\alpha = 10\,\text{deg}$. Figure 5.14(a) plots the localized movements of the free ends on the contact surface during the stick phase. The order of movement with respect to each other is illustrated in Fig. 5.14(b). If the direction of slide is from left to right (Fig. 5.14(a), inset), there will be two trends of slide propagation. The first is from the outermost point of contact (e.g., the $(-N)$-th and the N-th contacting points) to the innermost points ($(-N+1)$-th, $(N-1)$-th). The other slide propagation is from left to right ($(-N+1)$-th, $(-N+2)$-th, *etc.*). As a result, the last point of contact to move prior to a gross slide would be roughly in the middle of the second half of the contact line (Fig. 5.14(b)). There is a time delay t_d between the slide of the free end of the Great Beam and the gross sliding of the entire

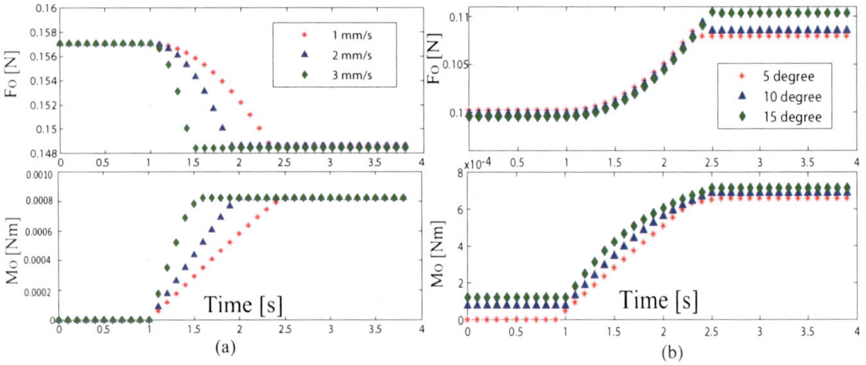

Fig. 5.16 (a) Plots of force F_O, and moment M_O on the fixed end of the Great Beam during the stick-slip phase along the (+) direction when α varies. (b) Plots of force F_O, and moment M_O on the fixed end of the Great Beam during stick-slip phase along (−) direction when the velocity of movement varies.

contact surface. In addition, the responses of F_O, and M_O are shown in Fig. 5.15(a). We found that during the stick phase the normal force decreases, while the tangential force increases, similar to results observed during the FEA simulation. The moment at which F_O becomes minimal is the start of the initial slide of the free end of the Great Beam. In contrast, the moment M_O becomes stable only when an overt slip of the entire fingertip occurs. The delay time t_d is perceived more clearly in this figure. Consequently, one might assess that the LDP dominates the transition from stick to slip phase, taking an important role in understanding the slip perception, and prediction of sensor's response.

A similar phenomenon occurs when the fingertip moves along the (+) direction, except that the normal force F_O increases, along with an increase in moment M_O (Fig. 5.15(b)). The delay time t_d is similar to that observed during the FEA simulation. Thus, by observing changes in force and moment at the fixed end of the Great Beam, we can determine not only the slippage of the fingertip, but also the direction of slide. Changes in F_O and M_O can occur when the tilting angle α is altered when sliding in the (+) direction, as in Fig. 5.16(a), or when the sliding velocity is altered along the (−) direction, as in Fig. 5.16(b). In these simulations, we can distinguish states of sliding, such as an earlier slip at higher velocity or a smaller tilting angle of contact.

Consequently, while the FEA simulation can lead to a more accurate but static estimation of stresses, dynamic analysis can reveal dynamic changes in force/moment. Both analyses agree that different states of contact can be reflected by the responses of a sensor located at point O. In particular, prior to the onset of incipient slip, the normal force at point O diminishes along the (−) direction or increases along the (+) direction, and the tangential force increases, resulting in a delay between the slide of the Great Beam's free end and the overt slippage of the entire contact surface. Both methods indicate the existence of the LDP during the transition from stick to slip. This suggests methods to detect the incipient slip of the sliding fingertip based on the responses of force/moment acting on the micro sensor. First, a change of M_O from kinetic to static may help determine the incipient slip, as previously proposed. Second, we can make decisions at the moment the value of F_O becomes minimal or maximal. By doing so, we can utilize the delay time t_d before the real incipient slip event actually happens, resulting in a high response for the controller. As a result, the LDP has helped us to predict the responses from the sensor, as well as to propose an appropriate way to detect the slippage of the soft tactile system.

5.3 Experiments on Incipient Slip Detection

We performed experiments to investigate the dynamic sliding motion of the tactile sensing soft fingertip, focusing particularly on incipient slip detection, and to show the system's potential in tactile and texture recognition.

5.3.1 Experiment Apparatus

To observe the sliding motion of a soft fingertip, we mounted it on a 2-DOF motorized linear stage consisting of two linear stages, an x-axis linear stage XMSG615, a z-axis linear stage ZMSG413 (Misumi, Japan) and a motorized stage controller MSCTL102 (Sugaru Seiki, Japan), which can precisely displace each axis with a resolution of $2\,\mu$m. The base of the soft fingertip was attached to a 3-DOF load cell PD3-32-10-105 (Nitta, Japan) to measure the total force/moment acting on the fingertip (Fig. 5.17, inset). The load cell-soft fingertip was subsequently mounted on the vertical z-axis stage, and made to slide on a fixed rigid surface plate along the horizontal x-axis (Fig. 5.17). The soft fingertip was controlled to slide on the rigid plate, the surface material of which can be changed. We hereafter call the coordinates attached to the sensor $O-xyz$; the axis of the linear stage controlled to move vertically the Z-shaft, and the horizontal axis the X-shaft. Signals from the sensor include F_z, M_x, and M_y, which are measured using a 12-bit ADC PCI card (CONTEC, Japan). The software, built in a Visual C++ 6.0 (Microsoft, USA) environment, can acquire data from the sensor, implement digital filtering algorithms, calculate, and output the commands to the linear stage driver at a highest frequency of 500 Hz.

5.3.2 Compression Test

Prior to conducting the experiments, we performed compression tests, using the INSTRON machine, to show the dynamic response of the sensor. In the

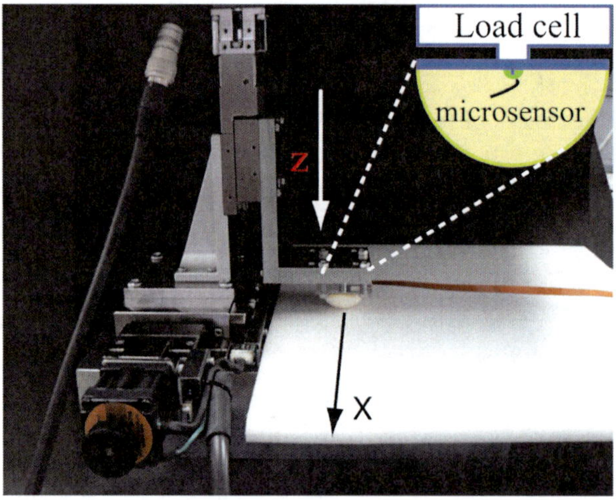

Fig. 5.17 Experiment apparatus. The soft fingertip is attached to the z-axis of the linear stage, and moved along the $z-$ and x-axes.

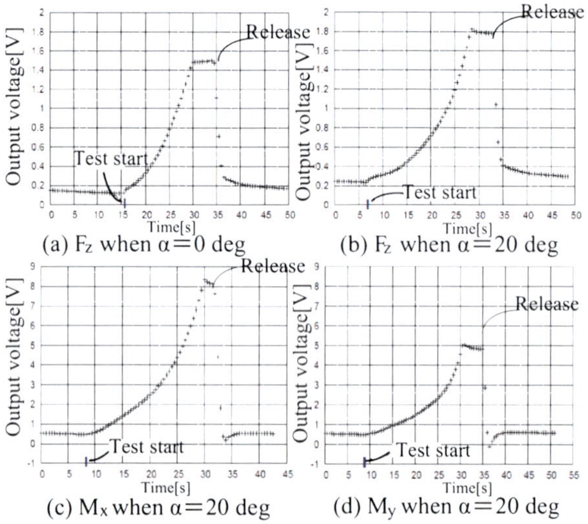

Fig. 5.18 Responses of sensor in compression test

tests, the head of the INSTRON was in contact with the soft fingertip, at a titling angle of 20 deg, until the load cell placed under the fingertip reached a predetermined value. In addition, the head was in contact with the rubber dome of the fingertip, at 60 deg from the x-axis. The responses (F_z, M_x, and M_y) of the sensor are plotted in Fig 5.18. We found that the value of F_z was larger when $\alpha = 20$ deg than when $\alpha = 0$ deg (Fig. 19a). The responses were sufficient to follow changes in the head of the INSTRON, such as acceleration

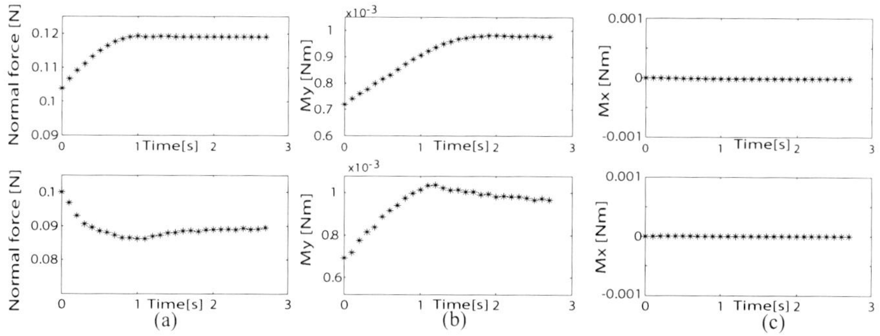

Fig. 5.19 Responses from sensor during stick-slip transition. The upper row shows responses relative to the (+) direction, and the lower row shows responses relative to the (−) direction.

of applied force or return. As shown in Fig. 19(b), the signal of M_x was larger than that of M_y, due to the head of the INSTRON being placed at a position tilted 60 deg from the x-axis [55]. Moreover, we found that, in all responses, signals passed across the initial value and returned to the initial value. As a result, sensor responses were fast enough to follow the change in deformation of the soft fingertip. We also specified that the maximum external force applied to the soft fingertip at which the sensor is broken was 21 N, or more intuitively with a contact depth larger than 5.1 mm.

5.3.3 Experimental Analysis of Sliding Motion of a Soft Fingertip

Sliding trials To create the same conditions as in the simulation, the soft fingertip was driven by the following process: during the first few seconds, the fingertip was pushed vertically onto the rigid surface at a contact depth d_n=2 mm; the fingertip was then moved horizontally along the x-axis at constant velocity v. The tilting angle $\alpha = 10$ deg. The coordinate attached to the soft fingertip was the same as that of the micro sensor. The dynamic sliding motion of the soft fingertip, including incipient slip, could be observed during this phase. This process was repeated during many trials to assess the consistency of the sensor outputs. Figure 5.19 illustrates the force/moment responses in voltage from the sensor in sliding along the (+) and (−) directions. Because the fingertip slides along the x-axis, the moment M_x is unchanged at zero value (Fig. 5.19(c)). Upon examining our experimental results in detail, we found that the soft fingertip had a normal force signal (F_z) and moment M_y, with an offset when it was pushed but had not yet moved. This offset could be caused by several reasons, such as an unbalanced Wheatstone bridge in the free state, bias from the sensor circuit, or a tiny asymmetry in placement of the sensor on the base. However, the main

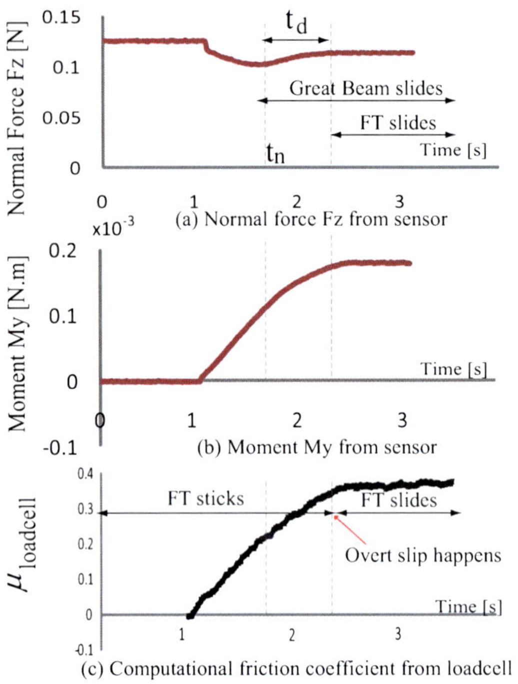

Fig. 5.20 Responses from the sensor and load cell during the stick-slip phase

reason was the normal force distribution when $\alpha \neq 0$, as explained in the above theoretical analysis. When the fingertip starts to move, slip has not yet occurred, causing deformation of the fingertip along the x-axis. During this phase, M_y increases for both cases (Fig. 5.19(b)). Interestingly, in Fig. 5.19(a), the value of the normal force F_z decreased markedly when moving along the $(-)$ direction and increases along the $(+)$ direction, as mentioned in Sections 5.2.1 and 5.2.2. F_z attains its minimum and maximum values right before M_y becomes stable, results consistent with those obtained during the dynamic simulation in Section 5.2.2.

Many methods have been used to detect the slippage of tactile sensing systems on the surface of objects. The most utilized method is the detection of sudden changes at the output of the kinetic coefficient of friction μ during the stick-slip phase [21]. This method is useful for point-to-surface contact or between rigid objects. The contact between the hemispherical soft fingertip and the object is surface to surface, with no sudden change in μ observed at the onset of slippage. This makes it challenging for the controller to determine the moment of slip in a timely and proper manner. Based on the simulation results using LDP, we can exploit a method to detect slippage based on observing the movements of the center zone, in which the change in normal

Table 5.2 Table of t_d Corresponding with Various Relative Velocities When $\alpha = 0\,\text{deg}$

Velocity [mm/s]	t_d [s]
0.5	3.2
1	2.0
2	0.7
3	0.43
4	0.31
5	0.2

force from the micro sensor can determine incipient slippage before it actually occurs.

Figure 5.20 shows responses from the sensor and reference load cell during the stick-slip phase in a special case when the tilting angle $\alpha = 0\,\text{deg}$. The kinetic coefficient of friction μ evaluated by the load cell plots two separate phases: stick and slip, and a smooth change between them. The signal of moment M_y has a tendency similar to that of the load cell, with a slight lag at the moment of slip. In contrast, the normal force is minimal at time t_n, or sooner than the moment of slip of the fingertip at time t_d (similar to that in Fig. 5.15). In this case, $t_d = 0.43\,\text{s}$. In fact, the normal force does not diminish markedly, with the gap between the minimum and maximum values being about 20 mN. Therefore, to detect the minimum value, the controller must continuously compare the value at each cycle to that during the previous cycle. Moreover, to assure that the minimum value is not a local minimum, the controller must continue for several cycles after t_n before judging. We found that about 10 samples, corresponding to 20 ms at a sampling frequency of 500 Hz, were sufficient. Using this method, we can determine the slip *prior* to its occurrence.

Responses from the sensor when some conditions vary are recorded in Fig. 5.21. When the velocity at which the fingertip is moved increases, slip occurs sooner, as illustrated in Fig. 5.21(a) and (b) for both directions of sliding. The moment of slip occurs more quickly along the $(-)$ than along the $(+)$ direction. This phenomenon can be explained based on the results of the FEA simulation in Section 5.2.1, in which the total normal force was larger when moving in the $(+)$ th an in the $(-)$ direction, entailing a larger tangential force, with the same coefficient of friction, to ensure slip, as stated by Coulomb's theory. This also causes a longer delay time t_d when sliding along the $(+)$ direction. Table 5.2 shows the changes in t_d corresponding to those of v; when $\alpha = 0\,\text{deg}$. The values in this table are averages from several trials. These results show that, the higher the relative velocity, the smaller the value of t_d. At $v = 5\,\text{mm/s}$, $t_d = 0.2$ efficient enough for the controller to properly assess the slippage of the soft fingertip. When v is small, t_d becomes larger, resulting in this method of detection being inefficient. Nevertheless,

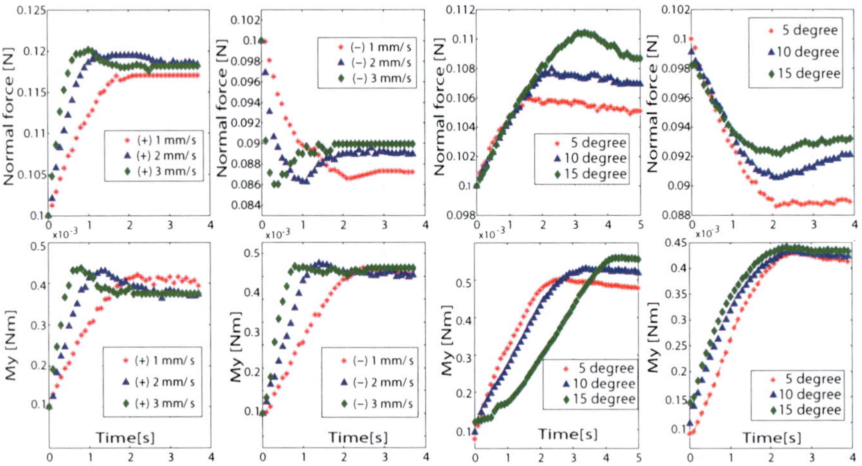

(a) Angle = 5 deg, speed changes sliding along direction (+)
(b) Angle = 5 deg, speed changes sliding along direction (−)
(c) Speed = 1mm/s, angle changes sliding along direction (+)
(d) Speed = 1mm/s, angle changes sliding along direction (−)

Fig. 5.21 Responses from the sensor in various slide conditions

in the object-grasping task, the object is accelerated by gravity, resulting in a high sliding velocity and making this method efficient.

When the tilting angle changes, results differ for the two cases of slide. When the fingertip moves along the (−) direction, the slip takes place earlier at a larger value of α; at a smaller value of α, however, the fingertip gives way earlier (Fig. 5.21(c) and (d)). Moreover, we were able to observe the offset values of the sensor when the angle varied, with a smaller normal force and a bigger value of moment at a larger angle, as described above in the section on dynamic analysis.

5.4 Concluding Remarks

This chapter described our initial research into a complete tactile-sensing fingertip that can be used in artificial robot hands during manipulation tasks, and use of the LDP in assessing slip perception for this system.

A MFMS, utilizing MEMS technology and piezoresistive effectiveness, was developed successfully. Fabrication and calibration processes were optimized and standardized, resulting in one of the first 4-DOF force/moment millimeter-scaled sensors in the world that is able to operate reliably and stably. The original sensor had a design that worked with 6-DOF in [52]-[53]. However, this design was complicated to fabricate, had larger dimensions, and greater crosstalk among signals. We sought to optimize the design and operation of a 4-DOF sensor, with one component of force and three components of moment, resulting in simplicity in manufacturing, simple measurement circuits, compact size, and high sensitivity. Moreover, we investigated the fingertip during bilateral sliding motion, not rolling or twisting; thus we only employed three components of force and moment. Depending on the purpose of future research, we can modify the design and operation of the sensor to better adapt to the application. This sensor, however, has a small critical fracture, which was 0.8 N under a normal load and 2 Nmm under a sheer load. Therefore, a hemispherical polyurethane rubber soft fingertip with a diameter of 20 mm was utilized to encapsulate the sensor, resulting in a complete tactile sensing fingertip. Using this soft fingertip, the system could flexibly handle heavier objects without concern of overloading and the fragility of the sensor. In addition, a data acquisition and processing structure, including hardware and software, was included to form an integrated soft tactile fingertip.

Since, as a result of using this design, the signals from the sensor were not the actual forces/moments acting on the fingertip, we performed careful mathematical and experimental analysis to show that, based on the responses of the sensor, many states of contact between the fingertip and the outside world could be assessed properly. Notably, the sensor can detect the sliding of the central zone of the contact surface, which is important in determining the point of fingertip slippage. Elaborating the transition from the sticking state to the slipping state of a soft fingertip has always been crucial in dexterous manipulation. Simulation and experimental results showed that the *center* zone is not the last to move before gross sliding of the contact surface, in point of fact, there was a delay. As a result, if we can perceive the slide of the *center* zone by some means, we will be able to assess the incipient slippage of the fingertip before the actual slippage. This delay time, t_d, is dependent on the sliding velocity of the fingertip. This strongly dues to the existence of LDP mentioned in the previous chapteres. In practical uses, such as grasping an object, the object is accelerated by gravity, causing a high relative velocity when sliding out of the fingertips. We therefore assessed sliding at high velocity. Although the exact detection time for robotics applications is difficult to perceive, it is sufficient for the sensors, which can emulate measured response times in humans (65 - 85 ms [96]). In contrast, determining the exact quantity of slippage for a sliding fingertip is challenging, making it important to determine the delay time between signals for the normal and tangential forces in order to ensure the timely detection of slippage. However, this delay time also varies as conditions change. Therefore, better detection requires a learning phase to be trained into this tactile system before use in each specific application. The quantity of slippage can therefore be determined using a parametric method to estimate its value when sliding conditions change. In this chapter, we focus on explaining the theoretical behavior of the fingertip when sliding and the ability to detect slippage with the assistance of the micro force/moment sensor. Further investigations have been performed, and will be published soon.

In this chapter, we have not dealt with contacts other than flat surfaces or fairly rough surfaces such as textiles. Analysis of complex surfaces can be performed using the static FEA method, but this is more challenging using dynamic methods. Contact planes with higher order equations include more complicated components (signals) of force/moment that will be exerted on the sensor. However, we believe it likely that a signal (at a specific frequency) will take charge of slip detection. This signal can be found using discrete wavelet transformation (DWT), which can generate both average and detailed signals in the time domain. We plan to employ this method to detect slippage between the fingertip and complex objects in the near future.

This chapter not only describes the development of a tactile system with a novel micro sensor, but also important investigations of the sliding motion of a soft fingertip. Several studies have assessed pushing and pure rolling of a soft fingertip, but none have explained how it slides or how the incipient slip occurs. To our knowledge, our dynamic analysis is the first to represent dynamically localized slips/movements acting on the contact surface, as well as to determine the delay time between the sliding of the center zone and the sliding of the entire contact surface. Robotic applications can employ our findings to develop tactile anthropomorphic fingers, which can detect slippage properly and in a timely manner. By introducing a fingertip with a small sensor, along with mathematical analyses, we have attempted to show that use of such a simple structure can recognize complicated behaviors of fingertip sliding. This should also encourage more research in the application of MEMS systems to robotics.

In conclusion, despite the remarkable reduction in the complexity of fabrication and signal processing, this tactile sensing system was found to be stable to operate, and sensitive to changes in contacts, thanks to the idea of BBM and LDP. This research will contribute significantly to the development of complete artificial robot hands. Future research will focus on assessing the ability of these tactile sensing fingertips to grasp objects stably, based on the capability to detect slippage as well as to control object posture. The objects will vary from rigid to deformable and be of many shapes. Moreover, new soft fingertips will be shaped similar to the human anatomy, employing many sensing points using micro sensors. This system will advance our ability to grasp and manipulate various objects.

6

Slip Perception via Soft Robotic Skin Made of Electroconductive Yarn

We have developed a fabric sensor knitted of tension-sensitive electro-conductive yarns. This sensor differs from other fabric-based sensors, in that the sensor itself is a piece of cloth with high density of knitted yarns. Each yarn has an elastic core, around which is wound two other separate, tension-sensitive electro-conductive threads, making this sensor *inherently* flexible and stretchable and allowing it to conform to any complicated surface on a robot, acting as a robotic skin. The pile-shaped surface of the sensor enhances its ability to detect tangential traction, while also enabling it to sense a normal load. Our aim is to utilize this sensor in applications involving relative sliding between its surface and a touched object, such as contact recognition, slip detection, and surface identification through a sliding motion. We carefully analyzed the static and dynamic characteristics of this sensor while varying the load and stretching force, to fully understand its response and determine its degree of flexibility and stretchability. We found that a Discrete Wavelet Transformation (DWT) may be used to indicate stick/slip states while the sensor is sliding over surfaces. This method was then utilized to detect slippage events acting on the sensor's surface, and to decode textures in a classification test using an Artificial Neural Network (ANN). Due to its flexibility and sensitivity, this sensor can be used widely as a robotic skin in humanoid robots.

6.1 Introduction

Recent robotics research on dexterous manipulation has sought to imitate the natural touch mechanism, as well as the anatomy of human fingers, to optimize the design of an ideal anthropomorphic artificial hand. This research has focused on fabricating tactile sensing systems that can convey information about contact force, friction, and roughness [[19]-[56]]. During stable grasping, a sensory system must also be able to recognize incipient

A.-V. Ho and S. Hirai, *Mechanics of Localized Slippage in Tactile Sensing*, Springer Tracts in Advanced Robotics 99,
DOI: 10.1007/978-3-319-04123-0_6, © Springer International Publishing Switzerland 2014

and overt slippage between the touching system and the object. Two methods are available to implement these requirements. First, as many necessary modalities as possible are incorporated into a single tactile cell. Recently, [51] and [56] described tactile modules designed to approximate multi-modal sensations. HEX-O-SKIN in [56] is equipped with multiple discrete sensors for temperature, acceleration, and proximity. Each taxel requires a separated data acquisition system with a microprocessor, then all of them must be communicated through a bus field. Also, the flexibility of the design is questionable. The second method focuses on enhancing single-modal sensors with intelligence for use in multi-purpose applications. For example, [57] added computer-vision techniques to a coarse tactile sensor for haptic exploration and object recognition. Our approach is similar to the latter, in that we sought to design simple sensors with multiple perceptions.

6.1.1 Fabric Sensors

There are several merits to using electroconductive yarns in sensor design, such as low profile, good dynamic response, and high durability. More information can be found in [58]. Some studies have used this material in the development of wearable health care systems for monitoring human ECGs (electrocardiograms) or breathing signals (see [59], [62]). In these applications, the yarns were stitched partly into everyday clothes at specific spots to form electrodes with which to detect changes in physical signs, or the kinetic movement of the human body. [60] developed a sensor in which hollow fibers were used to measure normal and tensional forces acting on wearable gloves. To the best of our knowledge, no such research has dealt with the detection of tangential traction, such as slippage on the surface. Conductive textiles were used to create artificial skin that can detect the location of contact using a reconstruction algorithm [[61] and [63]]. De Rossi and his colleagues have developed intelligent flexible textile based sensors to measure limb posture, utilizing multiple physical sensing systems, such as carbon elastomers and piezo-resistive effectiveness [[64] and [65]]. Nonetheless, this design was solely for the detection of curvature, and tactile information was ignored. A similar approach was used very recently by [66]. To our knowledge, however, there has been no research on sensors that detect tangential traction, such as slippage, on their surface. Therefore, our purpose was to create a low-cost, low-profile, but effective and accurate textile-based sensor that can detect slippage on its surface and that could be used in robotics applications.

6.1.2 Robotic Skins and Slip Detection Perception

Work on whole-body sensing by [67] involved the deposition of off-the-shelf infrared sensors on a large area of a flexible substrate. This design can supply proximity information on the surrounding environment in order to maintain safety during the interactions of humans and robots. However, tactile sensing cannot be obtained during these interactions. Alternatively, several types of tension-sensitive materials have been used recently to fabricate tactile robotic skins. [68] fabricated a tactile array skin that consists of two crossed layers of copper strips with a thin silicon rubber layer between them. with an applied force increasing the capacitance between the two strips. This sensor was used to detect contact location and object shape. However, there is no mention of slip detection, nor has texture recognition been reported. The authors in [69] utilize pressure-conductive rubber to detect slippage on a sensor surface. Experimental results show that when an object slides on the surface of a sensor, the rubber's resistance drops significantly. In addition, to discriminate between slippage and normal force change, the discrete wavelet method has been used efficiently. This research promises potential use in the detection of slip, but other abilities such as friction and texture perceptions have not been addressed yet. One commercial skin sensor is I- SCAN manufactured by NITTA Corp. which consists of a grid of tension-sensitive electroconductive ink lines (see [70]). The working principle is quite straightforward: when no load is present at an intersection, there is light contact between the inks, resulting in a high resistance between inks; when a force is applied to the sensing sheet, the conductive inks are pushed into a strong contact, causing the resistance to drop dramatically. However, this sensor is currently only used in static applications with respect to normal stress distribution, and not yet in dynamic ones such as slippage detection.

6.1.3 Texture Discrimination

There has been an increasing number of studies of late addressing surface identification utilizing machine learning methods (see [22], [71]-[73]). Authors in [22] used a neural network with several feature spaces, such as Fourier coefficients of the signal, minimum Euclidian distance from the mean spectrum, and high-level features such as friction, variance, kurtosis. Similar approaches are also found in [71]. Oddo *et al.* [72] exploited the k-nearest-neighbor classifier to encode roughness for discrimination of fine surfaces with 400-, 440-, and 480-μm spatial periods. It is difficult to say which method best performs, since this depends on many factors, from sensor design through to specific application. Most contacts in these researches were point-to-surface, reflecting detailed, unique information required for texture discrimination, such as the roughness and pitch of a surface. Our sensor always made surface-to-surface contact, thus most of the aforementioned methods were not suitable. Therefore, in this research, we employ several conventional methods for classification, from which we are able to find out the most suitable one.

6.1.4 Contributions

In this research, we describe a novel construction of an electro-conductive yarn and an application of a flexible sensor with slip perceptions. Instead of

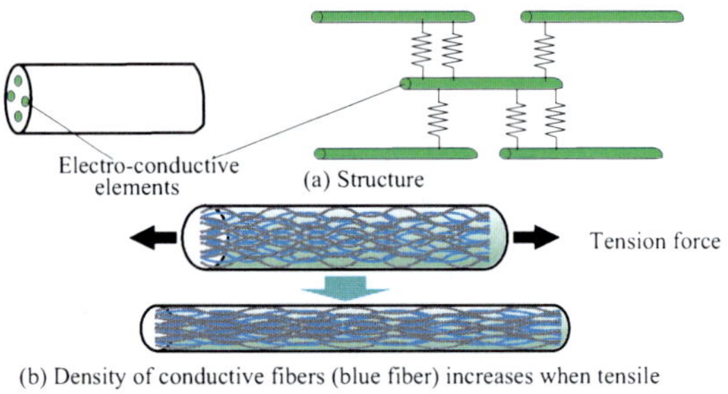

Fig. 6.1 Model of a tension-sensitive electroconductive yarn

using a common electro-conductive thread for knitting, we have proposed a *coupling design* of a yarn in which two electro-conductive threads were wound around an elastic core, permitting remarkable increase of resistance change over its stretched strain. Moreover, by utilizing these coupling yarns, we have introduced a novel design of knitting yarns with *continuous piles* arising on the surface, in order to complete a flexible/stretchable cloth sensor that is enhanced to detect *tangential traction*; which was, to best of our knowledge, never mentioned in previous research. In term of data processing, we have named a *slip indicator* for the sensor to identify/quantify the slippage acting on its surface by utilizing discrete wavelet transform (DWT). Moreover, we also have shown that for such surface-to-surface contact that indicated by simultaneous micro contacts, feature-based supervised learning extracted from DWT data is most suitable for the sensor to recognize/discriminate various textures. As a whole, this section is an example of how the localized displacement phenomenon (LDP) proposed in our previous research can contribute to the fabrication of novel sensor, as well as data processing, in order to recognize the slip actions. This is expected be a substance for the community to develop their own research.

6.2 Construction of a Slip Sensor

6.2.1 Tension-Sensitive Electroconductive Yarn: State of the Art

A tension-sensitive electroconductive yarn is a mixture of polyester fibers and stainless steel fibers that have high conductivity (see Fig. 6.1). This yarn is tension sensitive, *i.e.*, when a tension force is applied to the yarn, it is deformed laterally due to the elasticity of the polyester fiber. This yarn was invented by the KurodaShoji Company in Komatsu, Japan. As the density of the electroconductive elements increases, this causes the number of contact points among the conductive elements to change. As a result, the resistance of

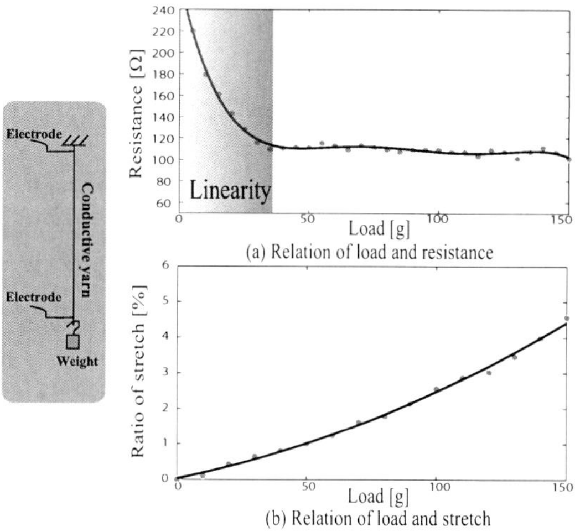

Fig. 6.2 Relation between resistance/stretch level over load in the tensile experiment. The inset pictograph illustrates the measuring setup/.

the yarn falls. Conversely, the smaller the tension force, the more the density of electroconductive elements decreases, resulting in an increase in resistance. An experiment was conducted to measure the change in resistance of one yarn, which was placed under tension by increasing tensile forces at one of its ends. Figure 6.3(a) shows the resistance-load relationship in the tensile experiment. Two distinct phases were recorded: a linear phase and a saturation phase. The linear phase, in which the resistance of the yarn changes significantly when the load increases, only occurs in a narrow load range (0 to 30 g). Beyond this value, the saturation phase occupies, in which the resistance value is stable even under high load.

Figure 6.3(b) shows the level of stretch over natural length as a percentage, and the applied load. In this graph, we can see that the linear range of the yarn, which is from 0 to 30 g, corresponds to less than 1 % of strain. Thus, this yarn only can be used in applications with a small load, and with a remarkably small rate of stretch. We cannot use this yarn for the application of slippage detection as this task requires a large degree of shear traction.

To increase the linear range, a double-coupling structure is employed. In this design, two yarns are wound around an elastic core made from polyurethane rubber, one over the other and they are wound in opposite directions, to eliminate the inherent twists of these yarns (see Fig. 6.3(a)). The merit of this structure, hereafter called *coupling yarn*, is that when it is stretched the electroconductive yarns become tensile, causing a decrease in the overall resistance. Thus, the elasticity and the tensile ability of the yarn increase remarkably. We then conducted an experiment (see Fig. 6.3(b)) to investigate the response of the yarn's resistance over the input stretch. The

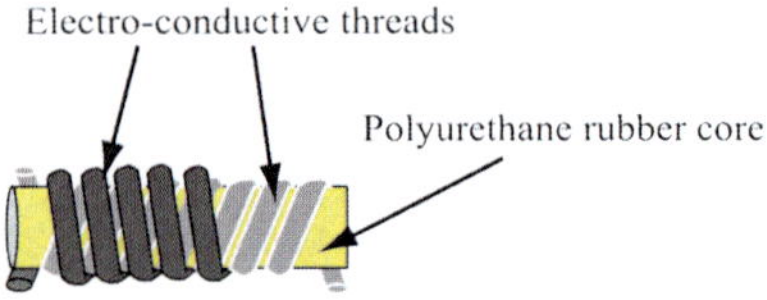

(a) Coupling yarn design

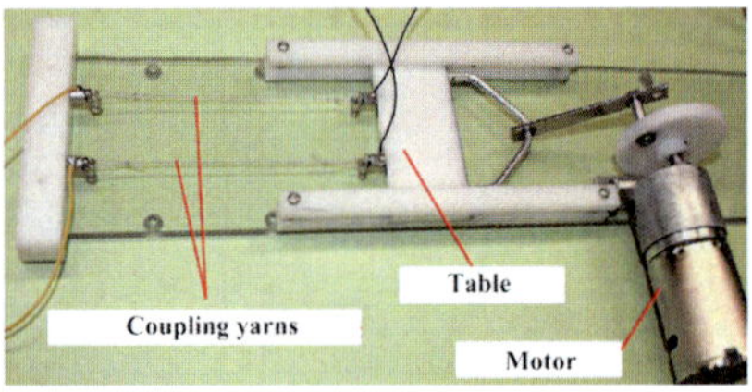

(b) Experimental setup

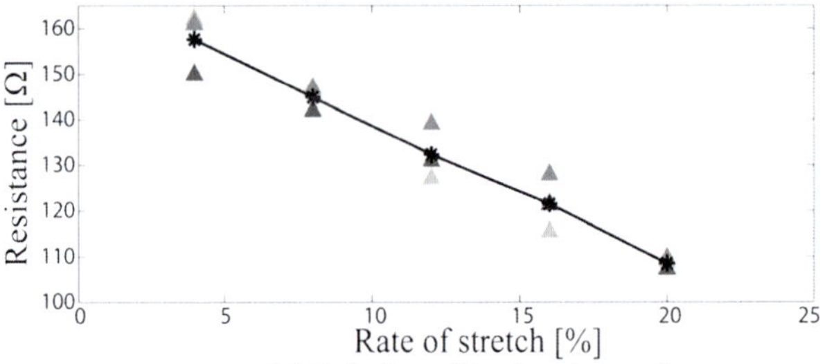

(c) Relation of load and stretch

Fig. 6.3 Relation between resistance/stretch level over load in the tensile experiment of the coupling yarn

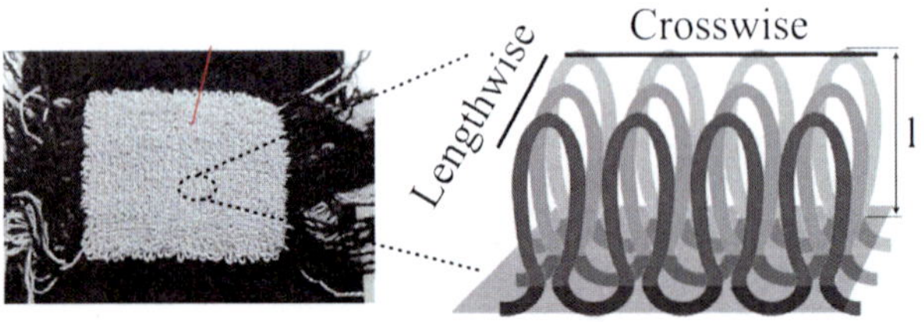

Fig. 6.4 Slip sensor with piles shaping the surface

relationship between the resistance of the coupling yarn and rate of stretch is illustrated in Fig. 6.3(c). The linear range is increased to up to 20 % of the tensile strain, significantly greater than observed for a single electroconductive yarn. This characteristic of coupling yarn enables its use in applications requiring high tension.

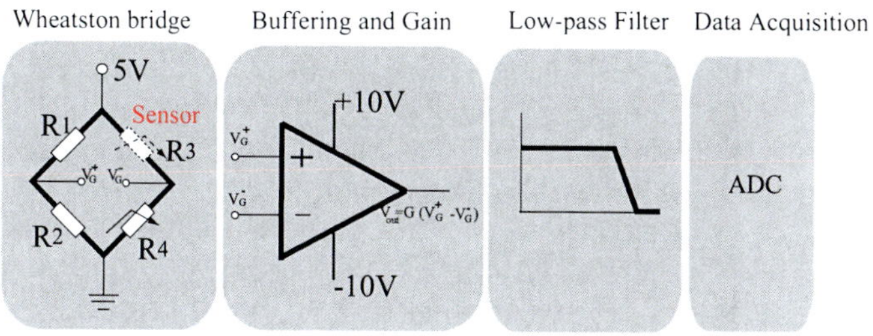

Fig. 6.5 Conditioning circuit diagrams for reading sensor's output

6.2.2 Slip Sensor

The design of the sensor was inspired by the Localized Displacement Phenomenon (see Part I for details) which dictates that, during the stick-to-slip phase of a sliding soft fingertip, micro movements take place on the contact surface during the transition, with overt slippage occurring only when all the contacting points give way. This phenomenon was derived from our proposed Beam Bundle Model (BBM) used for modeling sliding contact of soft objects. Thus, if these micro movements can be detected, either in a direct or *indirect* way, a method can be developed to determine how and when overt slippage occurs. The direct method can involve using cameras to track movements at each point, or by using a tactile array sensor that can dynamically sense trajectories of contacting points. The indirect method can take into account localized displacements into its transduced output. While direct methods can intuitively examine the sliding of fingertips, they are difficult to implement. In contrast, indirect methods are simpler to design, but require advanced data processing methods to accurately assess slippage.

To fabricate a fabric sensor that can detect slipping on its surface, we knitted double-coupling conductive yarns (as *warps*) in with the basal yarns (as *wefts*), which are neither tension sensitive nor conductive. However, if we were to employ the knitting method used in the fabrication of plain weaved fabrics or similar products in which the surface of the fabric is flat and smooth, the tension strain in the yarns caused by traction stress on the surface would be small, resulting in an insufficient output change in resistance. Therefore, to enhance the tension strain of the sensor, the yarns are weaved to form *pile cloth*, *i.e.* each yarn has many continuous loops that come up to surface level called *piles*. Figure 6.4 shows a complete fabric sensor measuring 5.0 cm in length and 3.5 cm in width. The length of each pile on the surface is less than 1.0 mm. These piles play an important role in detecting traction on the contact surface of the sensor. This sensor was fabricated by Okamoto Corp., Osaka, Japan.

The sensor was connected to one branch of a Wheatstone bridge circuit. A fine-tuning variable resistor (R_4) was bonded to the other branch in order to balance the bridge ($V_G = 0$) when no load was applied to the sensor. Thus, a change in the sensor's resistance would result a non-zero value of V_G. The V_G signal was then fed into an instrumentation amplifier (INA125, Analog Device, USA) with gain of G previously wired to a data acquisition system (AD12-16U(PCI)EV-Analog to Digital Converter, CONTEC, Japan). The sensor's output was filtered with a low-pass filter with a cut-off frequency of 1 KHz (see Fig. 6.5). The entire system was sampling at 100 Hz.

6.3 Sensor Model and Simulation Based on the Beam Bundle Model

In this section, we model the physical characteristic of the sensor so as to predict its response when an object slides on its surface. The BBM model is employed to model the sliding soft fingertip. Moreover, each contacting pile is modeled as a viscoelastic element, and its deformation during sliding of the fingertip is elaborated. Of particular note, we found that that movements of piles on the contact surface are important in assessing the incipient slip of the fingertip, which is similar to the LDP idea.

6.3.1 Modeling

As stated in the previous section, piles on the surface of a sensor play an important role in the sensor's sensitivity. Thus, we focus on analyzing the changes in pile morphology in order to extract electric change in the sensor. There are two phases of slide action that cause change at the output: the push phase and the slip phase. In this model, we assume that those phases happen sequentially (pushing and then sliding). Both cases cause stretch in each individual pile. The summation of the strain in all contacting piles is used to calculate appropriate electric change at the output of the sensor.

Each pile is considered to be constituted of a two perpendicularly connected *visco-elastic* elements, each includes of a pair of paralleled spring/damper (with stiffness k_i calculated by Young's modulus E, and viscosity c_i), due to the fact that the coupling yarn's core is polyurethane rubber (approximately visco-elastic). The normal element is taken in charge when the pile is deformed by vertical load, while the tangential element is activated when a tangential traction appears. Deformation of each pile will be represented by the deformation of the Voigt element. Estimation of the stiffness of each element can be calculated based on its geometrical shape, while the viscous moduli were chosen suitably. Moreover, since a pile is stitched at its two ends, thus the deformation of each pile does not depend on its neighboring piles. As a result, it is sufficient to analyze behavior of one pile, then synthesizing for all contacting piles of the sensor.

For sake of generality, we consider a contact of the sensor's surface and an object generates an irregular two-dimensional area on which the normal force distribution is non-uniform. Such contact can be seen widely in robotics when

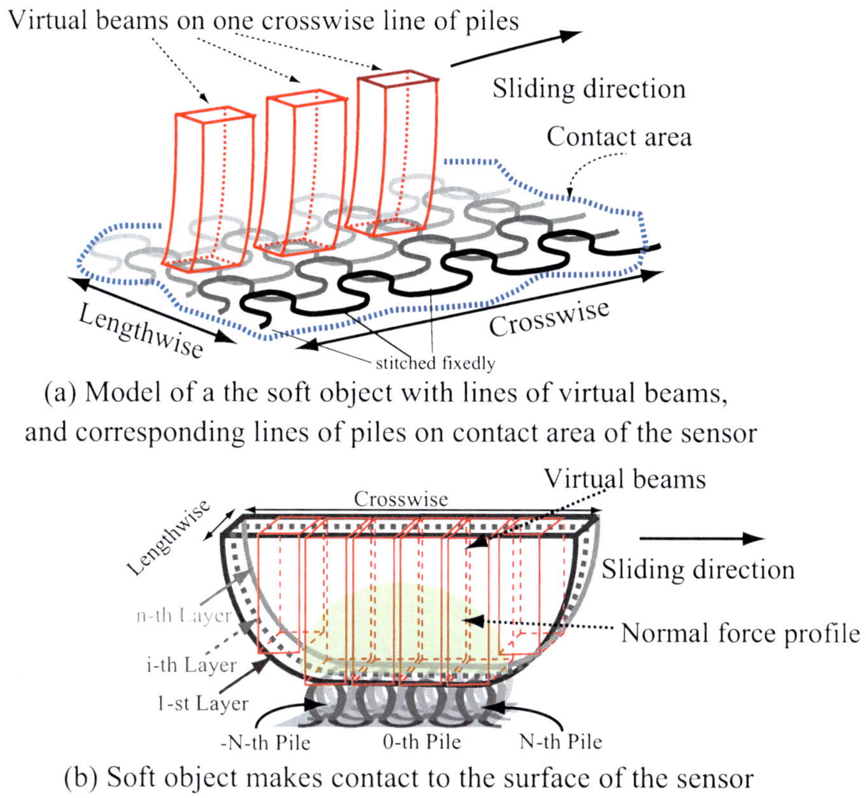

Fig. 6.6 Model of a soft object and sensor piles

a soft object with irregular outer shape (e.g. cylindrical and hemispherical soft fingertip) or non-homogenous structure (e.g. human fingertip) makes contact with a flat surface. In case of a rigid object, this distribution is consistent over the contact surface, thus it is easier to analyze compared to soft objects. By utilizing our proposed BBM, we assumed that the soft object is comprised of a finite number of virtual *soft beams* which are compressible and bendable, and possess different geometrical characteristics based on the object's shape (see the previous chapter) as illustrated in (Fig. 6.6(a)). Thus, a deformation of the fingertip while making contact is represented by the deformation of all these beams. Each *beam*'s end is assumed to make contact with one *pile* on the contact surface. Interaction of these two dominates stretching strain of the pile. These beams are aligned in multiple layers, each layer is corresponding to one crosswise line of piles as illustrated in (Fig. 6.6(a)). Therefore, it is sufficient to calculate strains of piles on one layer crosswise. On the other layers, it is supposed to happen similarly.

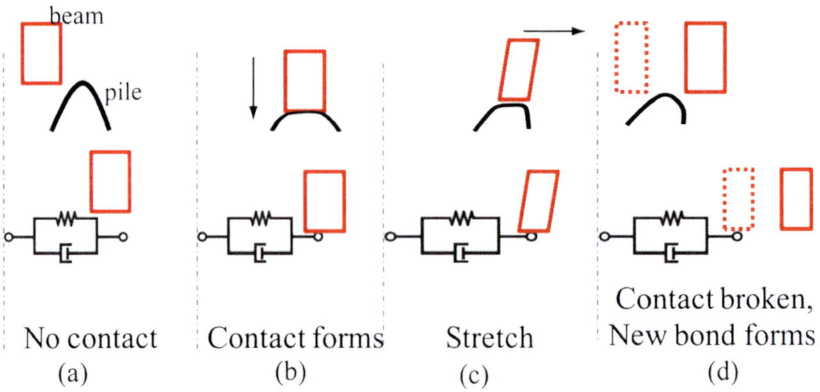

Fig. 6.7 Virtual layers of contact for modeling. The upper row shows the contact phases between a virtual beam and a pile; the lower row are appropriate physical models of those phases. First, the beam's end approaches and makes contact with a pile to form a bond (a) and (b). When the fingertip starts to slide, but the bond has not broken yet, the pile is stretched (c). When the bond is broken, the beam breaks away and makes contact with the next pile, while the current pile is occupied by the other beam.

Let us analyze the process of construction/destruction of a contacting bond between a beam's end and a pile (Fig. 6.7(a)-(d)). When a beam makes contact with a pile by a pushing action, the pile stretches horizontally, say Δx_i^n (Fig. 6.7(b)). When the SFT starts to slide, it deforms, causing the beams to bend gradually. Therefore, on the contact surface, slip has not happened yet (see Fig. 6.7(c)). In this phase, due to the bond, the pile is pulled by the beam's end, leading to the stretching strain Δx_i^t of the pile. The bond will be broken when $\Delta x_i = \Delta x_i^n + \Delta x_i^t$ reaches its critical value Δx_{limit}. This value depends on the characteristics of the pile, normal stress, and contact friction. Usually, it is determined based on the linear range of the yarn as mentioned in Section 6.2. When the bond is broken, the current beam will make contact with the next pile, while the current pile tends to bounce back to its original state. However, the neighboring beam will come into contact with and form a new bond with the current pile, causing this pile not to bounce back (see Fig. 6.7(d)). This process happens similarly on all contacting piles of the sensor. The equation motion for this process is summarized as follows:

$$F_{bending}^{i} - kx - d\dot{x} = \Delta m\ddot{x}, \tag{6.1}$$

where Δm is the reduced mass on the contact bond, and $F_{bending}^{i}$ is the bending force acting on the free end of the beam when it bends. This force differs for different beams, and the calculation can be referenced in Chapter 2 or 3.

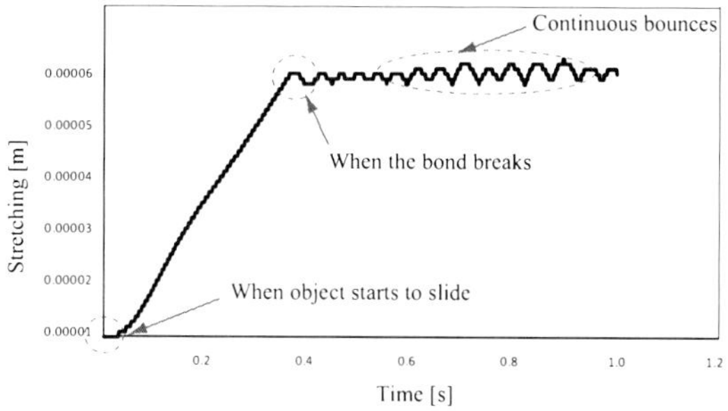

Fig. 6.8 Change in strain of one pile during the sliding motion

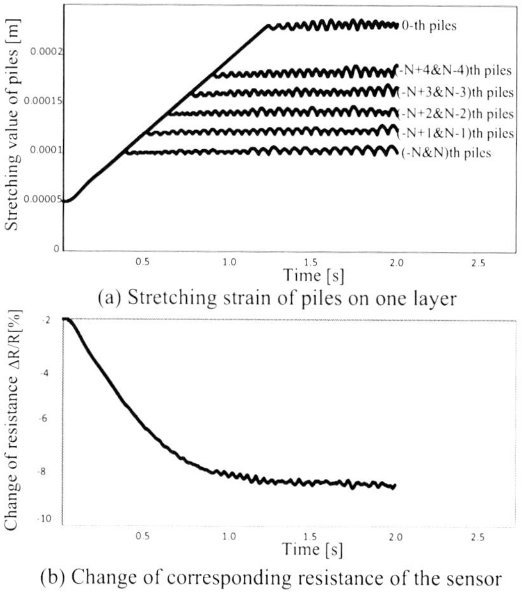

Fig. 6.9 Simulation results

6.3.2 Simulation

In this simulation, we chose the contacted object as a cylindrical soft fingertip (SFT). Profiles of beams on one layer (Fig. 6.6(b)), such as geometrical and material characteristics, are obtained in Chapter 2. With suitable parameters for simulation used in Chapter 2, we are able to assess the stretch strain of one pile during the process stated above. Because the SFT is hemi-cylindrical, the distribution of normal stress is non-uniform crosswise, but homogeneous lengthwise. Therefore, it is sufficient to calculate strains of piles on one layer crosswise. On the other layers, it is supposed to happen similarly (Fig. 6.6(a)).

In this case, the SFT is 10 mm in diameter, and 5 mm in width, thus there are 5 virtual layers of beam/pile along the lengthwise direction. Let us analyze movement of beams/piles on the arbitrary i-th layer. Assume that the SFT makes contact to the plane with contact depth of 2 mm, thus the number of contacting beams/piles for one layer is of $2N+1=11$, and these beams/piles are numbered from $-N$-th to N-th (see Fig. 6.6(b)). Young's modulus is $E=0.6$ MPa, viscous modulus $c=10$. Fig. 6.8 shows the resultant strain of one pile by solving Eq. 6.4. First, the SFT is pushed vertically on the surface of the sensor, causing an offset value of stretching strain due to normal strain Δx_i^n. Estimation of this value can be referred in the Appendix. Then, the fingertip is moved horizontally to create sliding motion. We can observe a steep increase when the beam sticks to the pile, causing this pile to be stretched. When the current bond breaks, a new bond forms, causing the strain of the pile to fluctuate around a stable value (see Fig. 6.8). For a layer of piles, the limit value Δx_{limit} differs depending on the relative contact positions with the beams. For the sake of simplicity, we assume that this value depends on normal force distribution of the SFT on the surface of the sensor; *i.e.* the higher the value of normal force, the higher the value of Δx_{limit}. As a result, we are able to assess the stretching strains of piles on one layer as plotted in Fig. 6.9(a). Simulation results of piles' stretching strain on one layer during stick-to-slip phase also reflect this phenomenon, showing that piles at the center of the contact area (e.g. 0-th pile) stretch the most compared to those at the boundary of the contact area (e.g. $\pm N$-th piles) before the bonds break (Fig. 6.9(a)), which obey the localized displacement phenomenon. In this configuration, there are 11 piles on one layer; and due to the symmetry of contact force distribution, there always exist two piles having similar behaviors during this process. In this figure, one can again observe the effect of localized displacements on understandings of how and when the slip occurs.

Because the SFT is 5 mm in thickness, the synthesized value of stretching strain on the overall contact surface of the sensor is obtained by simply multiplying the value in Fig. 6.9(a) to a gain of 5. One can see that the transition from the stick phase to the slip phase of the fingertip on the surface of the sensor obeys the LDP idea, whereby the localized displacements of piles decide how and when the slip happens, and so the corresponding output of the sensor. By employing the model and the LDP idea, we can ascertain the response of the sensor during slippage.

After assessing the change in strain of the piles, we synthesize the change in resistance of the sensor by exploiting the relationship between strain and resistance shown in Fig. 6.3(d). As a result, a sensor change in the sensor at the onset of the slippage can be detected, as shown in Fig. 6.9(b). One can observe that before the occurrence of the slip, the resistance of the sensor drops dramatically. This is caused by localized displacements on the contacting bonds on the contact surface between the fingertip and the sensor. After overt slippage occurs, the resistance remains unchanged due to the

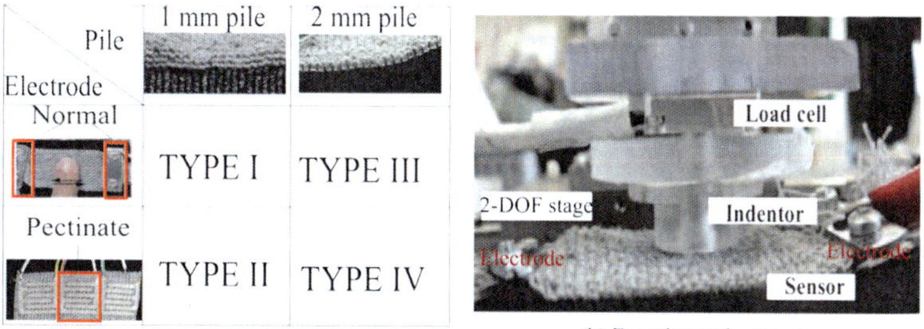

(a) Types of sensor used in this experiment (b) Experimental apparatus

Fig. 6.10 Configurations for the static test

continuous slip on the surface of the sensor. The steep of the resistance drop depends on the velocity, load condition, friction, *etc.*

Consequently, by conducting physical modeling of the sensor and implementing simulation in a scenario with a sliding soft fingertip, one can assess what happens on the contact surface and how the slip occurs. The LDP that we have proposed in previous chapters plays an important role in dealing with the modeling and simulation of this sensor. This model is not only useful in perceiving slip detection for the sensor, but also has the potential to be employed in a haptic environment in future work.

6.4 Characterization of the Flexible Sensor

6.4.1 Types of Sensors

We have employed two types of fabric sensors with 1- and 2 mm high piles, and two types of electrodes to connect to measuring circuits. In the first configuration, the two ends of yarn were clamped with two conductive electrodes. This is hereafter called *normal* electrode (see Fig. 6.10(a)). In this configuration, the transducing circuit will measure variable voltage across the entire fabric sensor, due to changes in resistance. Using this type of sensor,we were able to detect any load applied to its surface, regardless of the location of contact. The second configuration consisted of a *pectinate* electrode with two comb-like conductive poles arranged in alternate positions, allowing localized changes in resistance at the surface of the sensor to be measured. Each pectinate electrode was placed on the backside of the sensor, at which there were no contact piles; and the electrode size could be varied depending on desired resolution (see Fig. 6.10(a) or 6.32). We used a laser cutting machine to construct a thin plate with a pattern of comb-like poles. This plate was placed on the back side of the sensor, after which a suitable amount of conductive epoxy (Chemtronics, USA) was flattened onto it. After several hours, the plate was removed, and the shape of the pectinate electrode was attached to

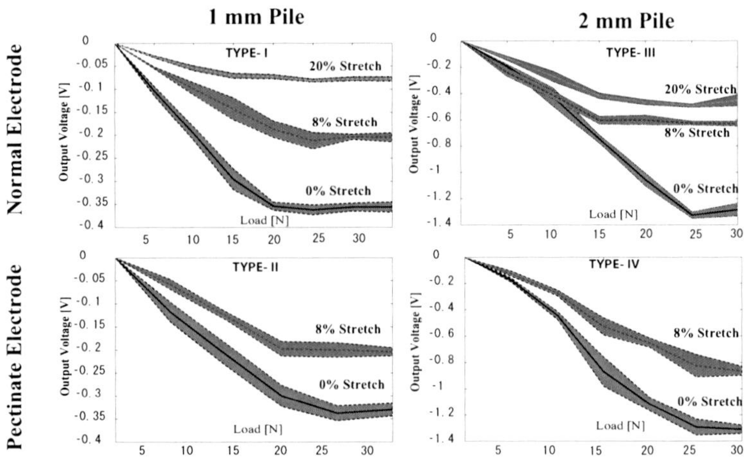

Fig. 6.11 Responses of four types of sensors in the static test. For each plot, a central line shows the mean value, and the outer band indicates standard deviation among ten trials.

the sensors, allowing us to measure the load acting on any location over the surface of the sensor. In short, we utilized four types of sensors, which are summarized in the inset table in Fig. 6.10.

6.4.2 Static Response Tests

Each type of sensor was tested by applying a series of static loads to its surface in the presence and absence of stretching strains. The sensor was stretched on a flat surface, such that stretch strain could be varied from 0 to 20 percent. A flat headed cylinder made of acrylic resin (MISUMI, Japan) and with a diameter of 20 mm was used to apply pressure to the sensor as an indentor. In order to apply normal pressure, the cylinder was attached onto a 2-DOF (degree of freedom) motorized programmable linear stage that allowed the cylinder to translate vertically and horizontally with a resolution of 2 μm. A 3-DOF loadcell (Tekgihan, Japan) was placed between the cylinder and the stage to measure the three components of force acting on the head of the former (see Fig. 6.10). The entire system was controlled and monitored by software written in Visual Studio C++, with a sample frequency of 100 Hz. An applied force profile of $\{0, 5, 10, 15, 20, 25, 30\}$ Newtons was followed. The stretched strain profiles (in percentages) were $\{0, 8, 20\}$ for Types I and II and $\{0, 8\}$ for Types III and IV. Pectinate electrode-based types could not be stretched much since the comb-like poles crack at high strain. Ten trials were conducted at each step for each type of sensor, and the means and standard deviations were calculated.

Fig. 6.11 shows the responses of the four types of sensor under static load and their stretched strains. The central line in each plot represents mean

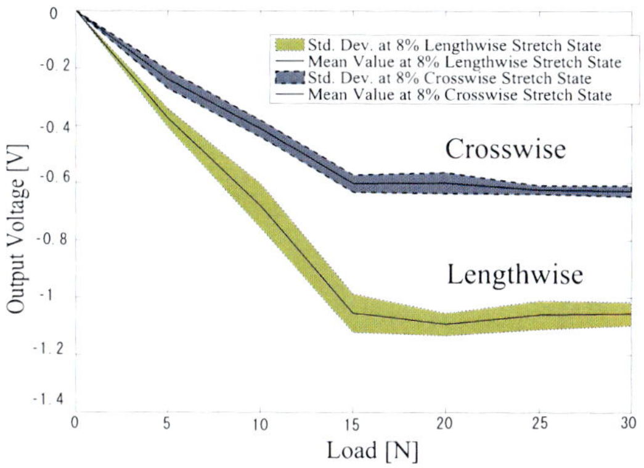

Fig. 6.12 Response of the sensor when it is stretched along crosswise and lengthwise directions

values, and the shaded band represents standard deviations at loading points. For all types of sensor at a stretch strain of 0 % (original state), the sensors responded linearly prior to becoming saturated at around 20-25 N of the applied load. Based on these plots, we were able to roughly *map* the change in sensor output as a function of applied load. Application to each sensor of a stretched strain less than 10 % resulted in linearity within 0-20 N, whereas the sensitivity (or derivative of each slope) decreased correspondingly. At a higher stretch strain (20 %), the 2 mm-pile sensor showed greater sensitivity than the 1 mm-pile type, indicating that, with 2 mm-high piles, the sensor can bear a larger range of stretch strain than the others. Moreover, stretching of the pectinate electrode within 10 % altered the sensitivity of these electrodes slightly when compared with normal electrode-based sensors. Consequently, with normal electrodes the sensor can bear a stretch strain of up to 20 % of its natural length, but may show low sensitivity. In contrast, the sensitivity of the pectinate electrode changed less when the sensor was stretched only within 10 % of strain. Although a 2 mm-pile sensor may produce a wider range of stretch strain than a 1 mm-pile sensor its response was less sensitive and stable, since the gain used in the instrumentation amplifier was much higher for Type III&IV than for Type I&II sensors (10 and 120, respectively).

In this experiment, the sensor was stretched in the crosswise direction (see Fig. 6.4) since yarns are knitted as warps. If the sensor is stretched lengthwise, only weft basal yarns deform, causing a small effect in conductive yarns. For example, in Fig. 6.12 two plots of 2 mm-pile sensors are shown, when it was stretched with 8 % strain in the lengthwise and crosswise directions, with the former having a sharper slope than the latter. Thus, for applications in which a sensor bears only a static load, we were able to calibrate the system it so that the sensor could automatically map changes in output voltage to applied force. With an applied force less than 20-25 N, at which the stretched

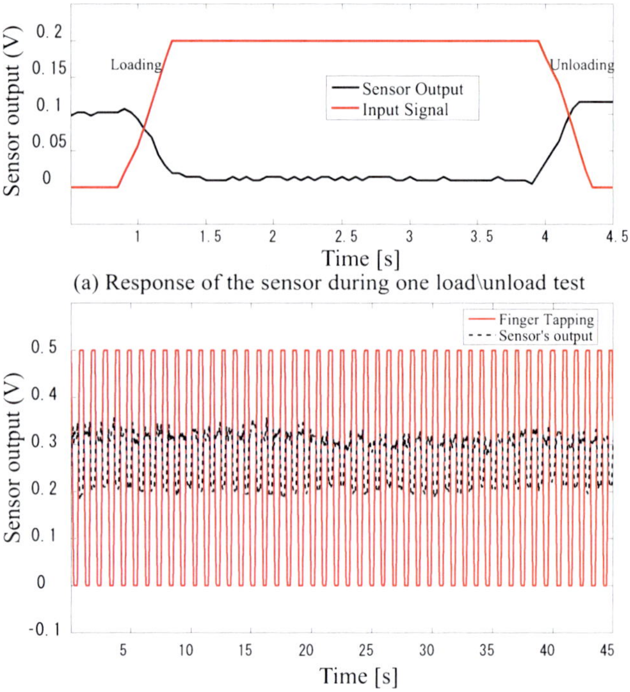

(a) Response of the sensor during one load\unload test

(b) Response of the sensor during continuous load\unload test

Fig. 6.13 Response of the sensor during loading/unloading tests

strain was within 20 % for Types-I&III and 10 % for Types-II&IV, the output voltage of the sensor is linearly correlated with the applied force with high confidence ($R^2 > 0.9$).

6.4.3 Dynamic Response Test

In this section, we investigate the sensor's ability to adapt to a dynamic load/unload test, to assess its hysteresis, and a sliding test to show its potential to detect dynamic changes in response to tangential traction. We utilized the apparatus used in the static test, with each trial repeated five times. Fig. 6.13(a) plots a response of the sensor output when a trapezoidal pulse applied to the surface of the sensor, showing that the sensor counters correspondingly to the slopes of the loading/unloading inputs. We also conducted a consecutive trapezoidal inputs with frequency of 0.8 Hz in 45 sec, and observed the stability/repeatability of the sensor's output as shown in Fig. 6.13(b). The upper bound's variance is more significant than that of the lower bound of the sensor output, due to the fact that piles takes time to recover to the original state because of the viscosity, and this recovery is slower than the input's frequency.

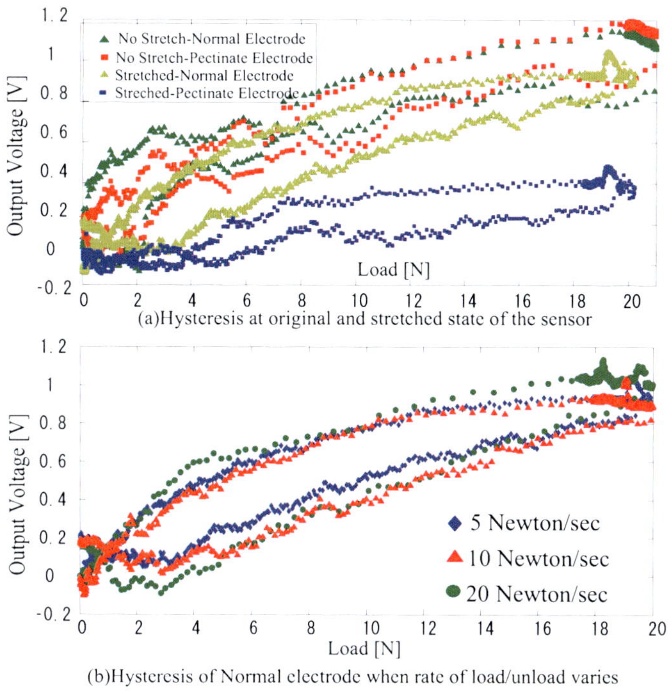

Fig. 6.14 Hysteresis of 2 mm-pile sensor Type-III&IV

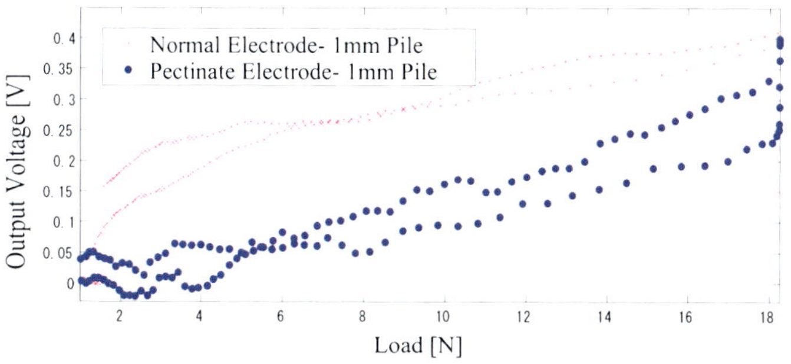

Fig. 6.15 Hysteresis of 1 mm-pile sensor Type-I&II

Hysteresis

In this test, we applied a normal force, increasing from 0 to 20 N (within the linear range) at various rates, then rested the cylinder at the maximum force at 3 sec, followed by unloading it at the same rate.

Fig. 6.14 shows the hysteresis of Type-III&IV sensors in their original state and in a 10 % stretched state. In the original state, the hysteresis generated by Types-III&IV did not differ greatly, although Type IV had a slightly smaller gap and better linearity at small loading/unloading force than Type-III (see Fig. 6.14(a)). When the sensor was stretched, its output produced less noise than in the original state since hysteresis plots were smoother. Nonetheless, the hysteresis of Type-III was larger than in its original state, whereas Type-IV's, although less sensitive, showed good hysteresis with a small gap and better linearity over the load-range. As a result, the fabric sensor is likely *more* reliable in terms of dynamic load/unload at small stretch strain. In this state, more unnecessary noise was eliminated by deformation of conductive yarns than at its original state. Consequently, even static tests can map from applied load to the sensor's output voltage in the linear range, the sensor is less efficient at mapping dynamic responses to continuous changes in load/unload. By changing the rate of applied load/unload, we investigated its dynamic responses (see Fig. 6.14(b)). The hysteresis gap widened when the rate of load/unload was increased, showing that its dynamic response decreased at a very high frequency externally applied force. The hysteresis of the sensor itself was due to the core of the yarn being elastic core (see Fig. 6.3), thus having intrinsic hysteresis and making that harder to deform when loaded than when unloaded [28]. Moreover, elastic hysteresis increased as the rate of loading/unloading increased, allowing the phenomenon illustrated in Fig. 6.14(b) to be assessed. Nonetheless, the observed hysteresis was lower for Types-I&II (see Fig. 6.15) than for Types-III&IV, since shorter piles have less elasticity than the higher piles.

Consequently, all types of sensors responded to continuous changes in dynamic loading/unloading within their linear range, showing that they can be used to estimate applied force during operation. Unlike conventional sensors (e.g. loadcells), we did not attempt to calculate the precise force acting on the surface of fiber sensors, since the output of the latter includes both hysteresis and noise. Human skin also cannot sense the exact force acting on it; rather, individuals utilize the deformation of the skin and stimulus signals from mechanoreceptors underneath the skin to assess how much force has been applied *qualitatively*. Similarly, fabric sensors show timely responses to dynamic changes in load and can be used estimate multiple *ranges* of applied loading/unloading.

Dynamic Sliding Test

Fabrics sensor could not only detect dynamic changes in loading/unloading force, but also respond in advance to tangential traction acting on its surface, thanks to the special design of the *piles*. A simulation in Section 6.3.2 had suggested a trend of change of the sensor output during a slip trial of a soft fingertip. We therefore conducted an experiment in which a soft fingertip (SFT) made by polyurethane rubber slides on the sensor surface given by the 2-DOF linear stage. Initially, when the SFT starts to move, its head still sticks to the surface, causing piles to be stretched. The stickier the contact,

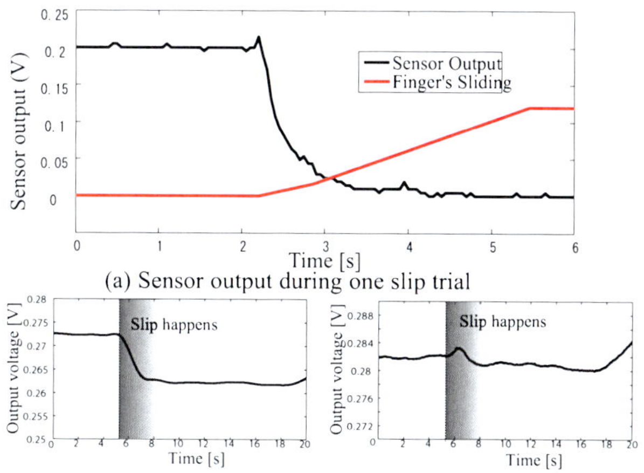

(a) Sensor output during one slip trial

(b) Slip along crosswise (left) and lengthwise (right) direction

Fig. 6.16 Experimental results when the cylindrical soft fingertip slips on the surface of the sensor

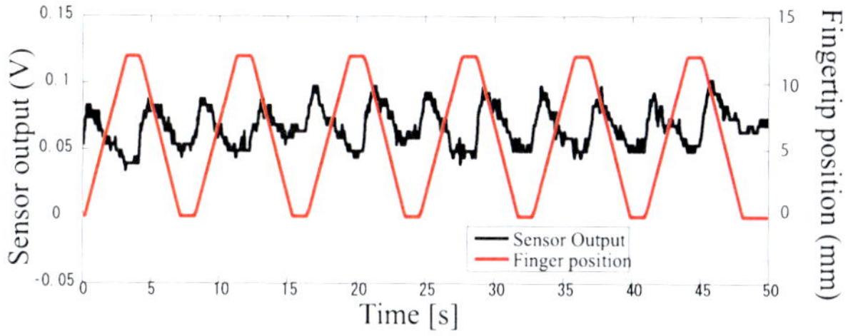

Fig. 6.17 Response of the sensor when the fingertip slides back and forward continuously

the more the piles stretch, reducing the overall resistance of piles. Fig. 6.16(a) shows the output of the sensor in one slippage trial. We observed a sudden change in output when the SFT started the incipient slippage on the contact surface, and a flatten output when the SFT slides overtly. Thus, by monitoring the change in resistance of the sensor, we can determine the sliding states of objects over the surface of the sensor. This experimental result is identical to the simulation result shown in Fig. 6.9(b), indicating that we are able to use the proposed model to predict the sensor output in term of sliding action. Moreover, we found that the direction of slide varied from lengthwise to crosswise (see Fig. 6.16(b)). A significant change was observed along the crosswise direction, with a lower change along the lengthwise direction. Thus, this sensor was more sensitive to the crosswise than to the lengthwise movement, although any arbitrary slip could be determined, regardless of the direction of sliding. We also conducted a durable test in which the SFT slides

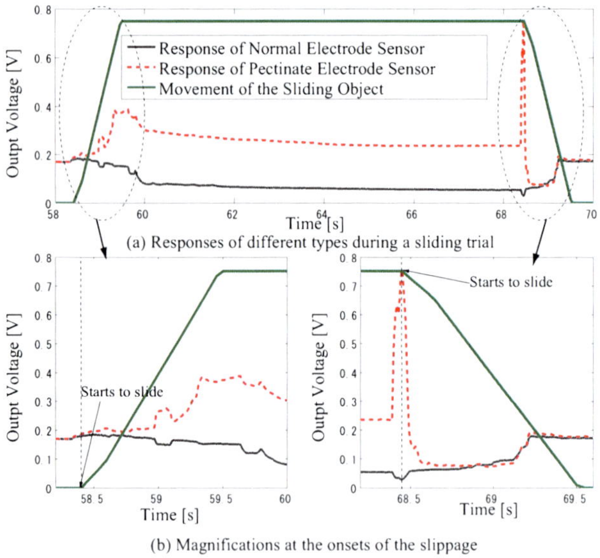

Fig. 6.18 Sensor's response during sliding of the object at original state (no stretch)

back and forward on the sensor's surface. Plots in Fig. 6.17 reveal that the sensor's dynamic response is equivalent to the sliding inputs.

We then tested the responses of our four types of sensors to sliding action on the surface. In this test, the cylinder was pushed and slid over the surface of sensors in one direction at a much faster speed of 6 mm/s, rested motionless, and then slid back to the original position. Fig. 6.18(a) illustrates two responses of Type-III&IV sensors and a ground truth showing the movement of the cylinder (measured by a laser displacement sensor, LB-01, Keyence, Japan) in the absence of stress (*i.e.* its original state). We observed fluctuations in sensor outputs while the cylinder moved over the surface. Moreover, by looking at two magnifications around two starting points of movement (Fig.6.18(b)), we found that the output of the sensor began to change right *before* that, thanks to the quick responses of the piles. This *lag* time, although, was important in enabling the controller to catch up to the object's slippage in a timely way. In considering the detection of incipient slippage in robotics ([19] and [54]), it is likely that the fabric sensor can be used to detect the relative movement of a grasped object or an interrelated action. The sensitivity of the sensor to slippage may change when the sensor is stretched. Similar types of plots in Fig. 6.19 reveal that the fluctuation in the responses of the sensors was less explicit compared with that observed in Fig. 6.18.

Consequently, the fabric sensor and its alternatives have been characterized fully using both static and dynamic methods. Depending on specific applications, each type of sensor can be chosen appropriately. For example, normal electrode-based sensors can operate better at highly stretched strains

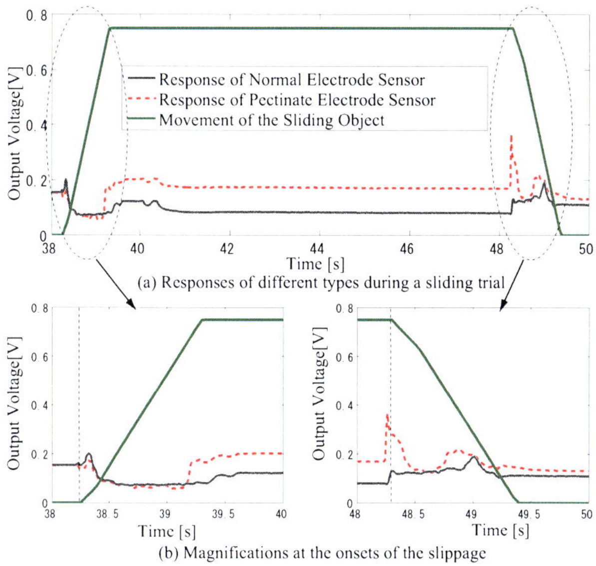

Fig. 6.19 Sensor's response during sliding of the object at stretched state

(over 20 %) than pectinate electrode-based sensors (less than 10 %). However, the latter respond in a more timely way to dynamic changes in load, both normal and tangential, at *localized position* over the entire surface of the sensor, which is promising in fabricating artificial skin with multiple sensing points. The slip perception of the sensor is unique compared with previous research, thanks to the performance of the *piles*. However, since both normal load and slippage both cause changes in sensor output, it is necessary to distinguish them, and more specifically, find a slip indicator for the sensor.

6.4.4 Discrete Wavelet Transform as a Slip Indicator

Slippage between an object and a sensor depends on many factors, including loading force, coefficient of friction, and speed of slide. Moreover, the output of the sensor is not always stable due to the characteristics of the cloth. Therefore, if the determination of slippage is based only on a change in output (Fig. 6.16), it will cause ambiguities. In looking at the continuous wavelet transform (CWT) of signals on a one slide trial in Fig. 6.20, we were able to assess the time-frequency representation of signals, allowing us to determine the dominant frequency at any point in time. In this analysis, we utilized a Mexican hat wavelet with width parameter equals to 1/16 (see [74]). Scales along the vertical axis in Fig. 6.20 represent the frequency, since the smaller the scale, the higher the frequency. Moreover, high frequency components appear at the moment of slippage, but *vanish* before and after the slippage.

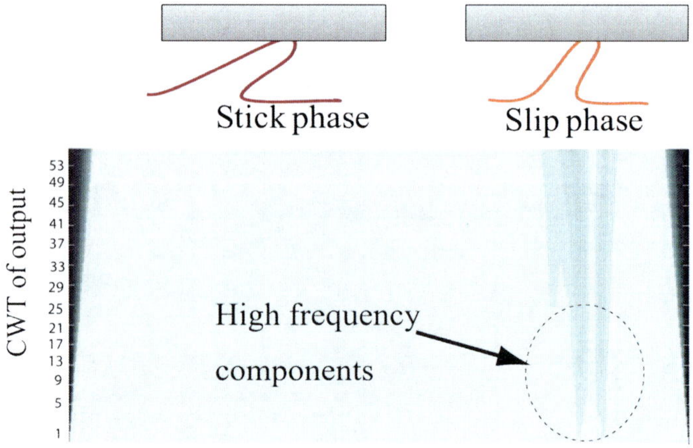

Fig. 6.20 CWT signal during one slide

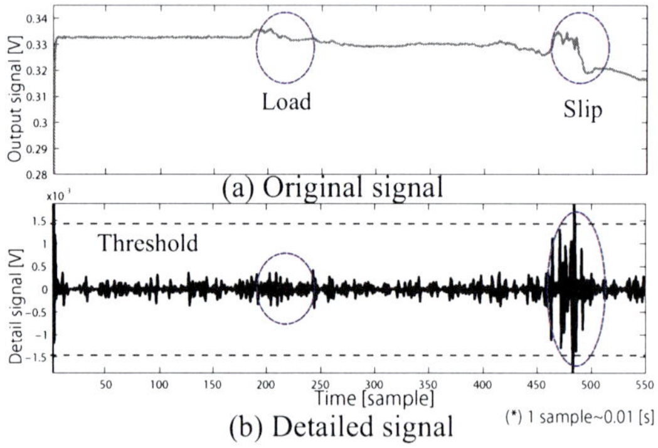

Fig. 6.21 DWT signal of one loading and sliding test

Thus, slippage of objects on the sensor can be determined by monitoring high frequency components of the sensor's output.

As described in the previous section, it is necessary to discriminate between load/unload and slippage states due to the similar signals of the sensor in response to these stimuli (downfall/uprise). As shown in Fig. 6.20, high frequency components can be observed at moments of slippage, but not during loading/unloading. Thus, the detection of high frequency components indicates that slippage is likely. Although Fourier transformation performs well in the frequency domain, it cannot accurately reveal the time it occurs. One of the major advantages of wavelets is their ability to analyze localized/small areas of a large signal, in which Fourier transformation cannot be used. Since wavelet analysis can extract information on discontinuities, breakdown points, and trends, this method is efficient in solving the above

Table 6.1 Slip Indicator Q_s for Slippage (in %)

1 mm pile			
No Stretch		Stretched	
Normal	Pectinate	Normal	Pectinate
0.089	0.0879	0.023	0.036
0.0178	0.0436	0.0119	0.0327
Normal	Pectinate	Normal	Pectinate
No Stretch		Stretched	
2 mm pile			

problem. We therefore used discrete wavelet transform (DWT) for a more efficient (fast) and accurate analysis. We fed the original signal passed through two complementary filters (low-pass and high-pass filters) and then merged into two signals. The high-scale, low-frequency component of the original signal is called the *approximation*, while the low-scale, high-frequency component is called the *detail*. We can therefore perceive both low- and high-pass filtered signals, especially the latter, showing a dominant frequency at one specific time [69]. In this wavelet analysis, we performed a single-level wavelet decomposition using the Haar wavelet (see [74]), then constructed approximations and details from the coefficients.

Fig. 6.21 shows graphs of the original and detail signals during load and slide tests. At the moment of load/unload, there were small changes (<0.0008 V) in the output of the detail signal. Interestingly, during slippage, there was significant fluctuation in the output of the detail signal, with an amplitude of around 0.0015 V (Fig. 6.21). Thus, the higher frequency component was more prevalent at the moment of slippage than during loading/unloading. By setting a proper threshold, as determined by the iterative method using many trials, the controller can easily distinguish between two ambiguous states ofloading/unloading and slippage, as well as detecting slippage. In summary, the DWT method is efficient not only in filtering the signal, but in extracting high-frequency elements over time, allowing the system to distinguish between loading/unloading and slippage during dynamic sliding tests.

We could also use DWT to quantify the energy generated by a slippage, allowing us to characterize the quality of the sliding. If $\mathbf{A}^\mathsf{T}$ and $\mathbf{D}^\mathsf{T}$ are vectors of *approximation* and *details* coefficients generated by passing a discrete signal with n samples through a DWT algorithm, then the quality of the slippage can be calculated as:

$$Q_s = \frac{\mathbf{D}^\mathsf{T}\mathbf{D}}{\mathbf{D}^\mathsf{T}\mathbf{D} + \mathbf{A}^\mathsf{T}\mathbf{A}} \times 100, \tag{6.2}$$

Table 6.2 Slip Indicator Q_s at Various Rates of Loading/Unloading

Load/Unload Rate [N/s]	0.81	0.75	0.610	0.520
Q_s [%]	0.80.002	0.70.0087	0.60.035	0.50.125

showing how much in percentage energy of slippage over energy of the signal. Using this equation, we could assess the slip perception of the sensor under various conditions of sliding contact, such as friction and velocity. Table 6.1 summarizes Q_s values as conditions change. The Q_s generated by pectinate electrode-based sensors was usually higher than that of normal electrode-based sensors. In addition, stretching of the sensors results in a smaller value of Q_s than in the original state with no stress. Moreover, sensors with higher piles are not more sensitive to slippage than sensors with shorter piles. The slip indicator was also calculated when the rate of loading/unloading varies, and summarized in Table 6.2. Looking at value of Q_s of Table 6.1&6.2, value of Q_s generated by the slippage is always higher than that of loading/unloading at moderate rate of force (less than 10 N/s). Nonetheless, with remarkably high loading/unloading rate (for example 20 N/s), high energy is still indicated by Q_s (Table 6.2). As a result, DWT can distinguish pushing and sliding, but not at a very high rate of load/unload changing.

6.5 Case Study I: Machine Learning for Texture Discrimination Based on Slip Action

In this section, in order to recognize different textures based on sensor data, we trained the sensing system by using several conventional machine-learning methods during slip motion on top of textured surfaces. Differing from conventional research, in which the contact between a sensor and a surface is point-to-surface contact (see [22], [72], and [73]), our sensor makes surface-to-surface contact, thus each localized contact of a *pile* on the contacting area contributes to the output of the sensor. This causes trouble and ambiguities during training and testing. Therefore, by conducting pattern recognition over several trials, we are able to find the most suitable one for our sensor. We demonstrated an experiment test bed in which the sensor was worn over a hemispherical rigid robotic fingertip that has a similar size to a human fingertip. The sensor is flexible enough to adapt nicely to the surface of the fingertip, acting as a skin to detect slippage of the fingertip (Fig. 6.22).

6.5.1 Data Acquisition

To collect texture data, we placed a skin-covered sensor atop a rigid hemispherical fingertip made of aluminum with a diameter of 10 mm, to emulate

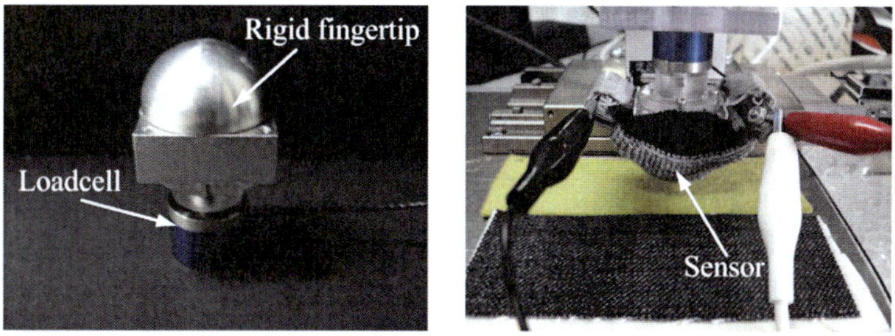

(a) Experimental setup for textures' data acquisition

(b) Textures used in the experiment and their assigned number

Fig. 6.22 Rigid robotic fingertip with the sensor covered in skin

the skin on a human fingertip. The fingertip was then mounted on a linear stage that can provide fine 3-D movement with a resolution of $2\,\mu$m (Fig. 6.22), and moved at a constant velocity of 5 mm/s over various textures, with a ramp-up time from zero is 0.5 s. Three textual surfaces were chosen for classification: denim, paper tape, and photo paper. These textures are different to each other, and generic for common objects in the environment with which a robot finger might interact. Moreover, paper tape and photo paper are both made from paper, thus the purpose of choosing them was to increase the difficulty of discrimination. Example data from the three textures are shown in Fig. 6.23.

Usually, when a person wants to assess texture of an object, they simply stroke a fingertip on the surface of an object, sliding it along one direction, and then lifts the fingertip. This action can be repeated several times if necessary, but not usually. In this experiment, the robotic fingertip copies this behavior when exploring a surface.

6.5.2 Naive Bayes Classification

In this section, we present a classification method exploiting a naive Bayesian classifier (see [76]), which has been shown to be highly efficient in discrimination tasks, even for the underlying Markov assumption that dominates the

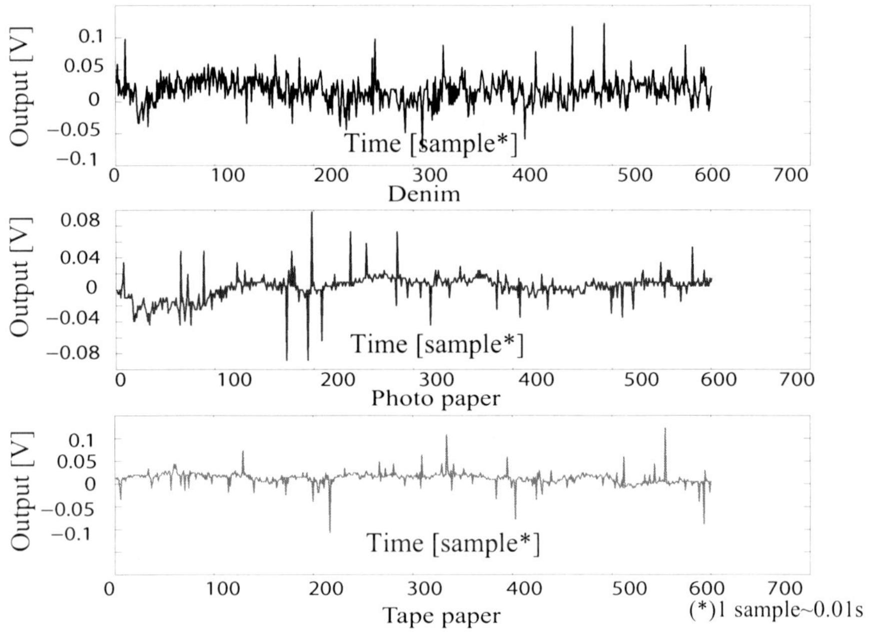

Fig. 6.23 Sample responses of the sensor when the fingertip slides on different textures

performance of this classifier (see [47]).The details are described below, and the validation results show that this method is reasonable in terms of both computation cost and reliability requirements.

Method

According to the Bayes theorem, the *posterior* probability distribution of state x given y can be calculated as:

$$p(x|y) = \frac{p(y|x)p(x)}{p(y)}, \tag{6.3}$$

where $p(x)$ is the *prior* probability distribution, and $p(y|x)$ is conditional probability along with the $p(x)$. According to the Bayes classifier, if the *posterior* conditional probabilities $p(x_i|y)$ are estimated over a set of $\{x_i\}$ given a single measurement y, the classifier will find the value x_m that has the maximum *posterior* probability:

$$x_m = \arg\max_{x_i} p(x_i|y) = \arg\max_{x_i} \frac{p(y|x_i)p(x_i)}{p(y)}. \tag{6.4}$$

As a result, if $\{x_i\}$ is a set of textures, then x_m will be the texture at which $p(x_i|y)$ is maximal. In the case of multiple measurements $\{y_t\}$, the arguments

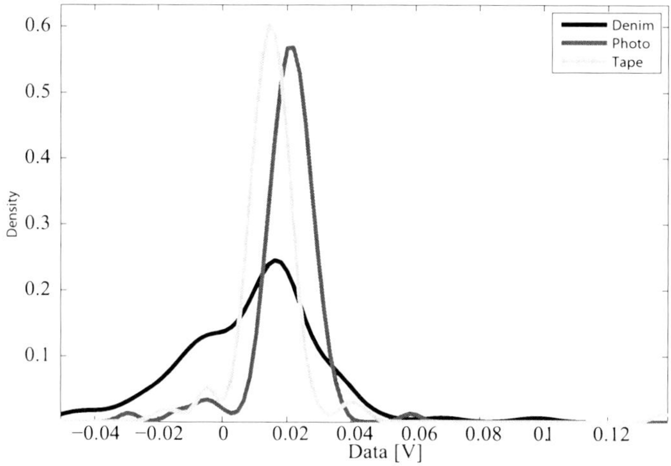

Fig. 6.24 Conditional probabilities of sensor output over textures. These distributions are estimated from the frequencies at which the values occur in training data, and a smoothed utilizing non-parametric fitting tool.

in Eq. 6.4 will be calculated in respect of a series of values instead of a single measurement:

$$x_m = \arg\max_{x_i} p(x_i|y_1, ..., y_t, ...y_n). \tag{6.5}$$

According to the naive Bayes classifier, there is an assumption, which is popularly known as the *Markov assumption*, that postulates that past and future data are independent, *i.e.* nothing *prior* to the current state can influence the stochastic revolution of the future states (see [47]). Then, the *posterior* conditional probability of x_i given the series of measurement $\{y_1...y_n\}$ will be replaced by the product of conditional probabilities for each measurement:

$$p(x_i|y_1, ..., y_t, ...y_n) = p(x_i|y_1)...p(x_i|y_t)...p(x_i|y_n). \tag{6.6}$$

Moreover, taking the logarithm on both sides of Eq. 6.6 does not affect the arg max in Eq. 6.5. Pooling all the hypothesis, the classification in Eq. 6.5 is estimated as follows:

$$x_m = \arg\max_{x_i} \sum_{t=1}^{n} \log p(x_i|y_t). \tag{6.7}$$

As a result, by finding the maximum argument of *posterior* probabilities, x_m will be the most likely texture given the series of measurements.

Implementation

To train the sensor, training data are taken from half of the measured data for textures, the other half later being used for validation. To determine the conditional probability of each measurement from the training data of one texture, the data should be made discrete, and the probability a single measured value y_t given texture x_i is specified basing on the number of appearances of the measurement value over the training data:

$$p(y_t|x_i) = \frac{n_{y_t}}{n}; \tag{6.8}$$

where n_{y_t} is the number of times that y_t occurs, and n is the total number of discretized values. Figure 6.24 shows the *prior* conditional probability distributions of the training data for the three textures. Each distribution was smoothed to obtain a nearly normal distribution. In Eq. 6.4, the probabilities of having a particular texture $p(x_i)$ are all equal to each other, *i.e.* 1/3. Because they are constant, they can be eliminated when estimating the maximum argument. The probabilities of measured values without given information about texture $p(y_t)$ are independent of the textures, thus it also can be eliminated as proposed in [76]. As a result, in consideration of the above arguments, Eq. 6.7 can be rewritten for convenient computation as:

$$x_m = \arg\max_{x_i} \sum_{t=1}^{n} \log p(y_t|x_i). \tag{6.9}$$

Validation

To validate the training system, we exploit three validation data sets obtained from texture data. Given data from an unknown texture, using Eq. 6.9 can help the system to classify which texture is most likely for these data. The details of classification and misclassification are shown in Table 6.3. One can see that denim classification outperforms the others with a high hit rate (80%), while paper tape and photo paper have similar hit rates (70%). On the other hand, the classification usually confuses paper tape and photo paper. If we look at the obtained probability distribution in Fig. 6.24, we can see similarities between photo paper and paper tape, while denim has a distinguishing distribution.

Consequently, by employing a naive Bayes classifier, we are able to classify textures using a trained sensor. This method is easy to implement, and is applicable in real-time classification. However, with the hypothesis of a naive Bayes classifier in which post and posterior state do not relate to the current state, this makes the accuracy of the classifier less persuasive. Therefore, other approaches are required that take into account the natural relation of measurements in a time series.

Table 6.3 Bayes Classifier Validation Data

Validation Data	Denim	Photo	Tape
Denim	80%	10%	10%
Photo Paper	0%	70%	30%
Paper tape	0%	30%	70%

6.5.3 Auto-regression Model and Artificial Neural Network

In this section we investigate the possible use of an autoregressive (AR) model to extract classifiable features of the sensor's output when exploring various textures. This model, in conjunction with machine-learning methods, has been exploited to classify human signals, such as electroencephalograms (EEG) of mental tasks (see [77], [78]). However, a way of exploiting this method of extracting representative features of synthesized data from sensors is a work still in progress. The fact is that a signal from a sensor is not always repeated, compared to an EEG, resulting in difficulties in modeling. In this application, a signal obtained from the sliding motion of a fingertip fluctuates around a mean value, as observed in Fig. 6.24. Thus, one might expect to model this response using an AR method.

Methodology

In statistics and signal processing, an AR model is linear prediction formula that is able to predict the output of a system based on previous outputs. Let AR(p) be an autoregressive model of order p, that is defined as follows:

$$X_t = c + \sum_{i=1}^{p} \phi_i X_{t-i} + \epsilon_t, \tag{6.10}$$

where ϕ_1 through ϕ_p are parameters of the model, ϵ_t is white noise, and c is a constant that is usually omitted. The parameters of the AR model can be estimated using several methods (see [77]). In this application, we use the Yule-Walker Estimator, which provides a route to estimation of parameters by replacing the theoretical covariances with estimated values.

In Eq. 6.10, p indicates the number of past samples used to represent the current sample, which can vary depending on how precise we would like the model to be represented. However, a large value of p is not coincident to high precision of the model. Thus, it is important to determine the order of the model that best fits the data. There is a popular criterion called the Akaike information criterion (AIC) which represents the trade-off between the estimated prediction error and the size of the model (see [77]). Consequently, by combining AIC and Yule-Walker estimator, we are able to represent the

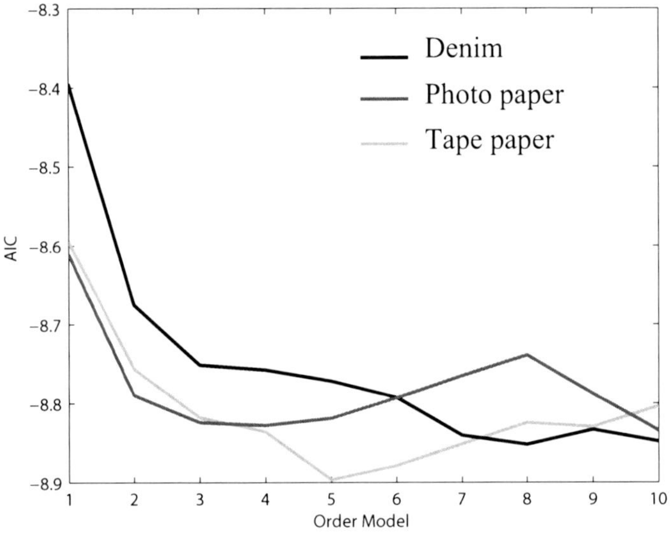

Fig. 6.25 AICs of the sensor's outputs over textures. The minimum value of each plot indicates the most suitable order for its AR model.

output of the sensor with high reliability, promising efficient features for learning/classification. These features are then exploited as inputs for training a multi-layer artificial network for classifying textures.

Implementation

In this application, we attempt to model a window of 100 samples of one texture's output. In this window, we then divide the time-series signal window into four segments, each of which contains 25 samples. Thus, to represent the window of data for one texture, a concatenated vector will be generated containing the parameters of all segments.

First, an appropriate order for the AR model need to be predetermined before conducting modeling. We used three windows of three textures to calculate their AICs at various orders. Figure 6.25 shows the plots of three AICs for model orders from one to ten. Each plot reaches its minimum: four for photo paper, five for paper tape, and eight for denim. As already mentioned, the minimum value of AIC often indicates the most sufficient order to fit the data. for the sake of simplicity, we chose the same order of six for the data of all textures, thus each model will be represented by a set of six parameters. The set of parameters of each segment is then estimated using the Yule-Walker estimator. The resultant sets of all segments of one window are concatenated into one vector. This 24×1vector represents the feature of one texture, and is usable in classification task. Figure. 6.26

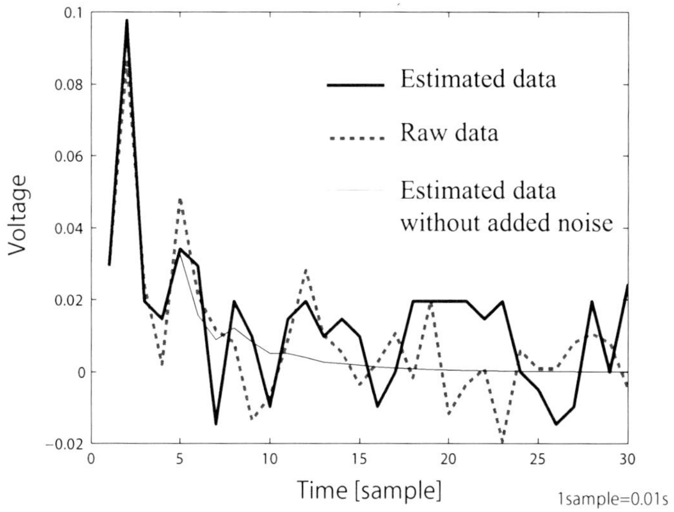

Fig. 6.26 Comparison between raw data and AR-based estimate

Table 6.4 AR-Based ANN Classifier Validation Data

Validation Data	Denim	Photo	Tape
Denim	67%	13%	17%
Photo Paper	17%	53%	20%
Paper tape	10%	30%	60%

shows the plots of raw data (dashed line), and the estimated line (solid line), which is constructed using a set of parameters to which are added white noise. One can assess the similarity of both responses, indicating that the AR-based estimation is sufficient to represent the output of the slip sensor.

After determination of the parameters, a feature vector of one texture is constructed by concatenating sets of parameters of segments in sequence. Three feature vectors of three textures afterward form an input vector of the ANN. As a result, each input vector has the dimension of 72×2, where the first column contains parameters, and the second one encloses flags indicating which parameter belongs to which texture. This input vector is then fed to the ANN for training or classification purposes. The ANN used in this application is a standard, feed-forward neural network, trained by an error back-propagation algorithm. Details of this network and its operation can be found in [80] and [81]. The number of layers was set at 3 as the classifier did not perform better when increasing layers. There are a total of 9 input vectors constructed from sliding trials over textures. In this application, we performed a cross-validation procedure to train the network. We first chose

70 % of the input data for training and 15 % for validation. After training and validation, the best network was fed by remained 15 % for testing phase. Table 6.4 indicates validation result. The mean successful rate is confirmed at 60 %, which is lower than that of naive Bayes classifier.

Consequently, even an AR model was found sufficient to model the outputs of the sensor over textures, but the corresponding classifier using ANN does not perform very well. This is due to the fact that estimated data without added white noise are close to zero over time, for all textures, as illustrated by the thin solid line in Fig. 6.26. As a result, this causes difficulty for classification requiring high confidence. Nonetheless, as this method is rather simple to implement, it is also promising for application as a primary classifier.

6.5.4 DWT and ANN Classification

The previous two classification methods have utilized raw sensor signals without a filter. One advantage of this is that details in the signal are not missed during analysis, enhancing the likelihood of successful classification. However, this causes difficulties in calculation (Bayes classifier) or modeling (AR-based ANN classifier). The success rates of these classifications were fairly high, but better classifiers are needed with more distinguishing features. One measure to achieve this is to exploit wavelet analysis, as mentioned in Section 6.4.4, more particularly the DWT method. Using this method, we are able to intensively examine the details of the signal, with findings promising for potential classifications.

Implementation

First, details of sensor's outputs over three textures are extracted using the DWT method, as shown in Fig. 6.27. These spectrums are quite distinguishable from each another. Moreover, next-to histograms show the distribution of signal over the measured data. One can see that the fitted probability distributions are similar to normal distributions, and vary clearly over textures. As a result, if features are extracted from these DWT analyses, one might expect classifiable feature vectors of textures for training and classification.

We employed a feature vector of five features extracted from DWT details in the time domain, which are:

1. Mean a.
2. Variance δ.
3. Standard deviation σ.
4. Entropy H.
5. Energy e.

While the first three features represent directly probability distribution of the DWT of textures, the last two indicate the characteristics of the wavelet. The entropy H reflects the degree of non-stationary or uncertainty that the

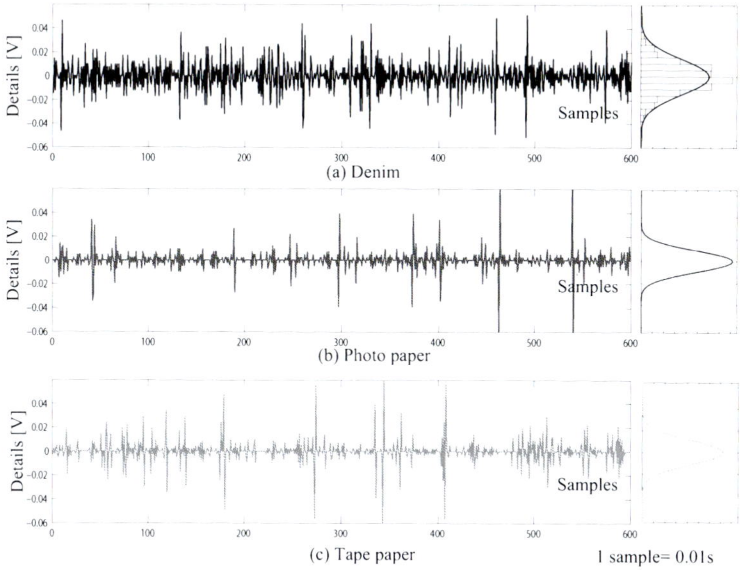

Fig. 6.27 Details of the sensor's outputs over textures extracted by using the DWT method. The next-to plot show the histogram of measurement values over their ranges. The estimated probability distributions is close to normal distribution. Similar results are also seen for photo paper and paper tape.

Fig. 6.28 Feature extraction process for one texture

variable possesses. Data with a high variance, *i.e.* a rough texture has high entropy, and vice versa. The entropy H for discrete random variable X is defined as [48]:

$$H(X) = -\sum_{i=1} P(X = a_i) \log P(X = a_i), \tag{6.11}$$

where a_i are possible values of X. The energy of the wavelet brings information to characterize signal energy distribution at different frequencies. A signal containing high-frequency components, which are considered important in assessing slip action as mentioned in a previous section on slip detection, possesses higher energy. Thus, a rougher texture, which causes more stick-slip phenomena when the sensor slides on it, results in high energy. For the sake of simplicity, in this application, we only calculate the total energy at different decomposition levels for classification. One might utilize energy at

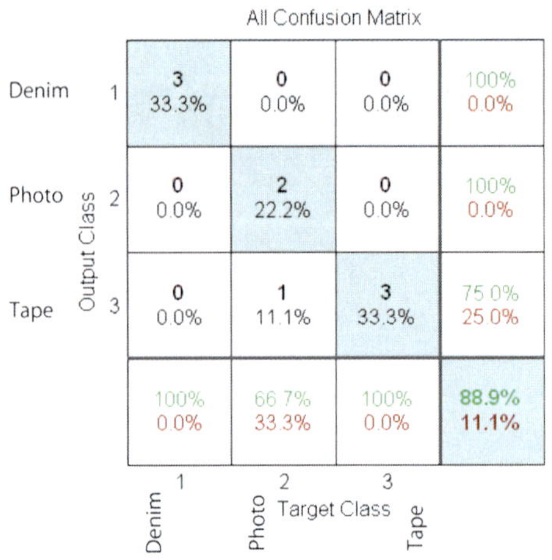

Fig. 6.29 DWT-based classification confusion matrix

specific levels, but this requires predetermined knowledge as to which levels are suitable. the energy at each decomposition level is defined as [79]:

$$E_j = \sum_k |d_{j,k}|^2, \ \ j = 1, ..., N, \tag{6.12}$$

where $d_{j,k}$ are wavelet coefficients, and N is the level. Then, the total energy of the signal after wavelet decomposition is:

$$e = \sum_{j=1}^{N} E_j. \tag{6.13}$$

In this application, N was chosen as 1, as a higher level did not perform any better. The summarized process for extraction of texture's features can be seen in Fig. 6.28.

Similarly to the above section, an input vector is formed by concatenating five features of each data window's DWT, and feeding it to a multi-layer feed-forward ANN. There are 9 input vectors, each measuring 5×2. We also performed cross validation during training and testing. The classification performance is shown in Fig. 6.29. One can observe that the DWT-based method gives better classification results than an AR-based one, with a hit rate of approximately 90%. The denim classification still outperforms the others, in addition to which the misclassification between tape and photo papers decreases significantly compared to the above methods. Consequently, the DWT-based ANN classifier can be said to outperform the other methods mentioned above, with a mean hit rate of 90 %. By exploiting features extracted from DWT, we are able to incorporate dynamic changes of stick-slip phenomena that feature a surface's texture into a classifier, resulting in a high success rate. Nonetheless, this method is quite expensive in computational terms compared to the aforementioned method. As a result, this method is suitable for use as a secondary classifier that requires high performance.

We then increased the number of textures by using this method for discrimination test for further investigation. Eight textual surfaces (see Fig. 6.22(b)) were chosen, including tile (numbered as 1), sand paper (2), photo paper (3), patterned board with holes (4), denim (5), alumina board (6), copper board (7), and patterned acrilic (8). Each of these textures differed from one other in term of material and texture, but were somewhat similar to increase the difficulty of discrimination, and those textures are sufficiently common in the environment for the robot finger to interact with them. For each texture, we collected 18 set of data, each consists of 6 sec window of time-serial data.

We still selected 3-layered ANN as a classifier. There were 144 input vectors, each contains five features. In this data-set, 70 % is for training, and 30 % is for validation. We performed a cross-validation process during training and testing. We conducted the texture discrimination on two types of sensor: 1 mm-pile and 2 mm-pile sensor, and three groups of sliding speed: 6 mm/s, 3 mm/s, and 1.5 mm/s. Six confusion matrices with successful rates of predicted textures and overall successful rates are illustrated in Fig. 6.30. General speaking, fabric sensor with shorter piles has performed better than the sensor with higher piles, due to the fact that the higher piles are not more sensitive than the shorter piles as aforementioned in Section 6.4.2. Moreover, it is likely a trend for all types of sensor that higher speeds of sliding generate better overall successful rates of prediction. Clearly patterned acrilic pad always generates very high prediction rate, almost 100 % for trials. For other patterned textures like sand paperback (2) or plastic pad with holes (4), successful predictions always outperform the other textures. Photo paper (3) that possesses very high friction can be predicted moderately well in case of the 1-mm pile sensor. Also, denim (5) can be easily recognized when conducted with the 1-mm pile sensor, while poorly discriminated at low sliding speed. In term of materials, one can observe that metal textures such as alumina (6) and copper (7) that do not possess patterned texture and low friction surprisingly generates very good predictions compared to other similar textures such as tile (1) or photo paper (3) even with high friction. It is due to the fact the fabric sensor is conductive, thus when making contact to conductive surfaces (metals) with different conductivity its characteristic may change correspondingly. Misclassification was caused among similar texture such as tile (1) alumina (6) and copper (7). As a result, the fabric sensor with shorter piles and at moderately high sliding speed could produce good discrimination result during supervised learning over patterned texture and metal materials.

Consequently, the DWT-based ANN classifier can be used for discrimination of textures with high successful rate. Using features extracted from

1mm-pile Sensor — v = 6mm/s (rows: Targeted Value; columns: Predicted Value)

	1	2	3	4	5	6	7	8
1	90	0	0	0	0	0	0	0
2	0	77	5	0	0	0	0	0
3	10	11	85	8	0	0	6	0
4	0	0	0	73	0	0	0	0
5	0	12	5	19	100	0	0	0
6	0	0	0	0	0	80	5	0
7	0	0	5	0	0	20	89	0
8	0	0	0	0	0	0	0	100

Overall: 86.5

1mm-pile Sensor — v = 3mm/s (rows: Targeted Value; columns: Predicted Value)

	1	2	3	4	5	6	7	8
1	68	0	0	0	0	0	14	0
2	0	93	6	0	11	0	0	0
3	0	7	73	0	13	27	0	0
4	0	0	0	86	13	0	0	0
5	0	0	12	14	63	0	0	0
6	0	0	9	0	0	73	0	0
7	32	0	0	0	0	0	86	0
8	0	0	0	0	0	0	0	100

Overall: 80.1

1mm-pile Sensor — v = 1.5mm/s (rows: Targeted Value; columns: Predicted Value)

	1	2	3	4	5	6	7	8
1	37	0	0	0	0	0	0	0
2	0	81	25	0	0	0	0	0
3	12	0	75	0	0	0	0	0
4	6.5	7	0	68	31	0	0	0
5	6.5	12	0	16	56	13	0	0
6	24	0	0	16	13	87	0	0
7	0	0	0	0	0	0	100	0
8	14	0	0	0	0	0	0	100

Overall: 75.0

2mm-pile Sensor — v = 6mm/s (rows: Targeted Value; columns: Predicted Value)

	1	2	3	4	5	6	7	8
1	76	0	0	0	0	10	17	0
2	8	61	25	0	0	0	0	0
3	0	0	65	0	0	0	0	0
4	6	8	0	58	31	0	0	0
5	0	31	10	16	61	13	0	0
6	0	0	0	26	8	77	0	0
7	10	0	0	0	0	0	83	0
8	0	0	0	0	0	0	0	100

Overall: 72.5

2mm-pile Sensor — v = 3mm/s (rows: Targeted Value; columns: Predicted Value)

	1	2	3	4	5	6	7	8
1	66	0	9	34	0	0	49	0
2	0	66	25	0	0	0	0	0
3	34	0	33	8	25	0	0	0
4	0	25	18	58	0	0	0	0
5	0	9	15	0	70	0	9	0
6	0	0	0	0	5	100	0	0
7	0	0	0	0	0	0	42	0
8	0	0	0	0	0	0	0	100

Overall: 66.3

2mm-pile Sensor — v = 1.5mm/s (rows: Targeted Value; columns: Predicted Value)

	1	2	3	4	5	6	7	8
1	72.5	12.5	0	6.5	0	0	0	0
2	0	75	6.5	0	15	0	0	0
3	12.5	0	31.5	0	16	0	0	0
4	6.5	0	18.5	50	0	0	0	0
5	8.5	12.5	18.5	18.5	56	10	0	13
6	0	0	0	0	13	90	0	0
7	0	0	12.5	6.5	0	0	100	0
8	0	0	12.5	18.5	0	0	0	87

Overall: 69.5

Fig. 6.30 Confusion matrices at different sliding speed and types of sensor. Numbered texture can be referred to Fig. 6.22. The green shading squares indicates successful rate of prediction of individual texture. The pink shading squares indicate the overall successful rate for each confusion matrix.

DWT, we were able to incorporate dynamic changes in stick-slip phenomena, featuring surfaces' texture into classifiers, resulting in a high success rate.

6.6 Case Study II: Human Fingertip Slip Detection

The slip between an object and the sensor depends on many factors, such as loading force, friction coefficient, and speed of slide. Moreover, the output of the sensor is not always stable due to the characteristic of the cloth. Therefore, if the determination of slip is based solely on the change of output, this will cause ambiguities. For example, with a rubbing action of a human finger on the surface of the sensor, *i.e.* changing the direction of slip continuously and periodically, the output changes continually but unclearly. As a result, it is necessary to use appropriate techniques to assure the determinations. The reason we chose the rubbing action is because this action is a critical one, one in which slips happen quickly, and the direction of slide changes continuously. If the sensor can detect this rubbing action, it will work well in other common slip actions. We have known that at the moment of slip, high-frequency components appear, but vanish before and after the slippage. As a result, the slippage of the objects on the sensor can be judged based on observing the high-frequency components of the sensor's output.

Let u_i and u_{i-1} be the discrete output voltage values of the sensor at moment t and $(t - \Delta t)$, respectively, with Δt being sampling time. Therefore, the derivative of the output at moment t, say d_i, is calculated as:

$$d_i = \frac{u_i - u_{i-1}}{\Delta t}. \tag{6.14}$$

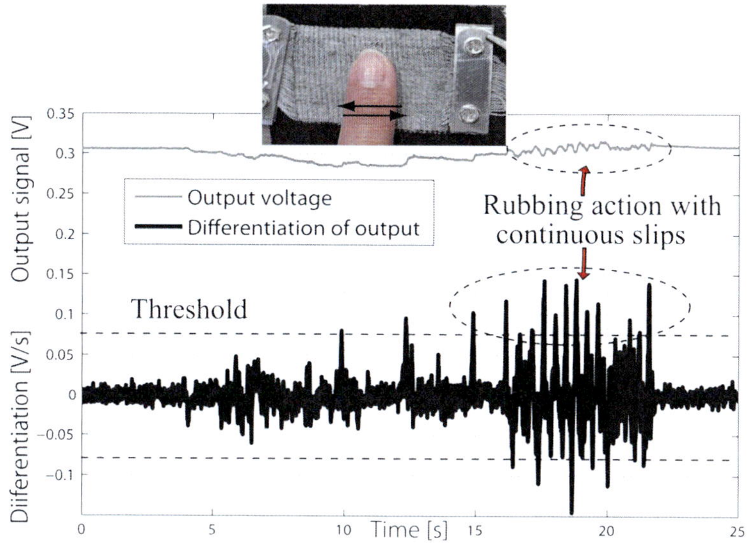

Fig. 6.31 Output of the sensor and its derivative during rubbing action

This can be considered as the stress rate of the shear traction over the surface of the sensor. By using this processed signal, sudden changes in the output voltage of the sensor can be perceived easily via peaks. As we can see in Fig. 6.31, whenever the human finger reverses the slide direction, that is, incipient slip between the finger and the surface of the sensor occurs, a corresponding peak appears in the stress rate d of the output voltage. The higher the rate of the incipient slip, the higher the peak. Therefore, if we can choose a suitable absolute value for threshold θ, slip detection can be judged if the value of d is over the threshold θ. To do so, a training data set $\{P\}$ containing values of derivative signal at moments of slippage is acquired. These data were collected while varying the contact conditions between an object (such as a human fingertip) and the sensor, such as load, velocity of slide. Based on this data set, threshold θ can be obtained appropriately by employing an iterative method.

We conducted an experiment in which four subjects were asked to rub their fingers at will on the surface of the sensor. They were not instructed how to rub it, for example: the necessary loading stress and rubbing speed. Threshold θ_h for the human fingertip was determined beforehand using training data $\{P_h\}$. A Microsoft Visual C++ program was used to process the output voltage of the sensor using a derivative method, determine whenever slip had occurred, and display the judgment in real time on the computer screen by turning an digital indicator on or off. The operation performed by each object was filmed and investigated after each trial. The rate between the number of slips detected by the program and that of one object averaged 78 %.

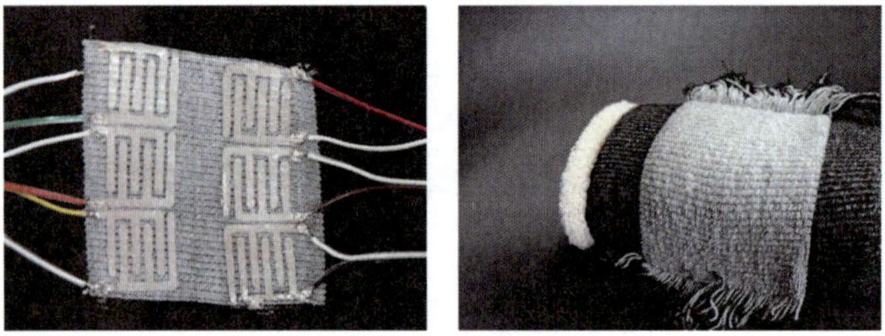

Fig. 6.32 Multi-sensing points on one piece of fabric sensor

By means of interviews and analyzing the movement of each object's finger after the experiment, we realized that the sensor is more sensitive to action at a fairly high rubbing speed, and a loading stress that is relatively high. The sensor is specially insensitive to an action with a very high loading stress or very low speed of slide. The reason for this stems from the quite small range of linearity of the sensor, and the ambiguity of the controller between the loading state and the slip state.

One of issues that needs to be considered is the threshold for processed data to determine the occurrence of slip. As we stated above, this threshold varies substantially when the contact state between object and the sensor changes. Therefore, to apply this sensor in actual robotic applications, we need to train the system with experimental data. Classification algorithms are possible choices for this purpose. The other issue is the direction recognition of the sensor. The sensor is more sensitive to motions in the crosswise direction than the lengthwise direction. By using data-processing methods, we are capable of detecting the slip regardless of the direction of slide on the surface. Nevertheless, some applications may require the recognition of slip direction of the sensor. The current design of the sensor can be considered as a 1-DOF sensor, which is most suitable for the crosswise direction. An improvement can be implemented to make the sensor become a 2-DOF one by stitch-separated yarns along the lengthwise direction. As a result, by synthesizing signals from crosswise and lengthwise yarns, we can distinguish the direction of slip of objects on the surface of the sensor.

6.7 Case Study III: Multiple Sensing Points with Pectinate Electrode

As described, pectinate electrodes can measure changes in resistance at localized locations on a surface. As a preliminary approach, we attempted to fabricate a piece of fabric sensor (2 mm-high pile type) with several sensing points. We deposited six pectinate electrodes on the backside of a 6 cm × 6 cm-sensor to construct a 3×2-array of sensing points with a resolution of 2 cm.

Each sensing point is hereafter called as a *taxel* (tactile element). The system was scanned at a sample rate of 100 Hz. It was then stretched onto a curved surface mimicking a robot arm (see Fig. 6.32) and tested in interactions with humans while pushing and sliding. Wiring systems are running under the hood inside one cable, to the outside data acquisition system.

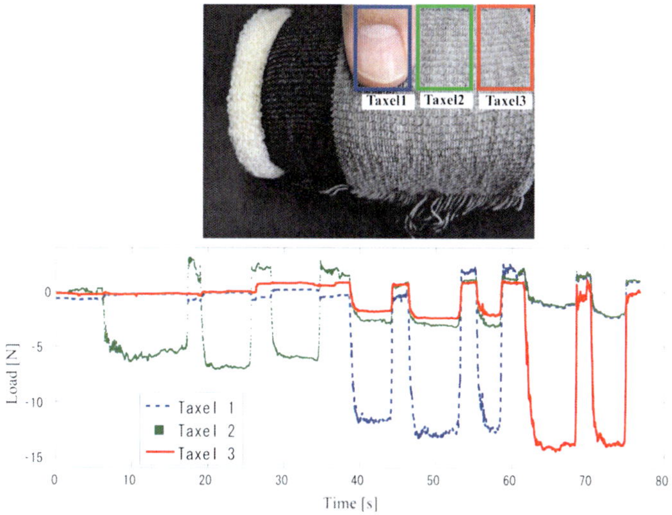

Fig. 6.33 Responses of taxels when a human fingertip pushes on them

6.7.1 Pushing

To test the ability of the sensor to detect different contact locations, the sensor pad was pushed at different locations. Fig. 6.33 shows the response of three sensing points on one row when pushed by a human fingertip three times each, in the sequence shown in the inset. By examining the responses of sensing elements, we could specify the point of contact. Each element responded quickly to a random load of human touch. Although there were cross talks among neighboring elements, the actually touched element generates the most significant signal, enabling the controller to specify the contact location. Crosstalk is only obvious with a high load of touching, *i.e.* more than 15 N as observed in Fig.6.33. One can observe some positive jumps at the sensor elements' output. In fact, at human fingertip when touching a conductive object like this sensor there exists a parasitic capacitor which might charge/discharge in combination with the sensor, causing these changes at the output of the sensor.

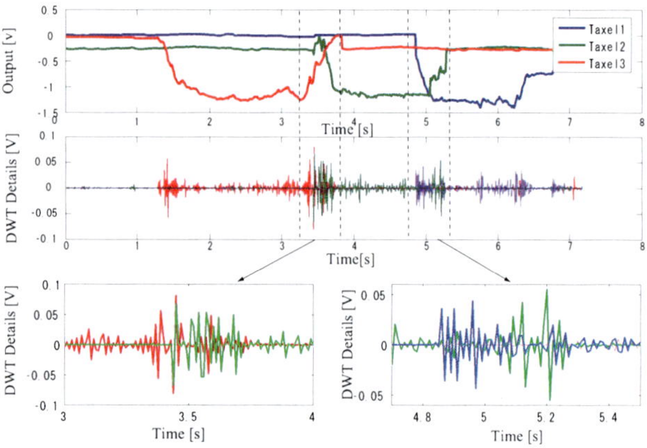

Fig. 6.34 Responses of taxels during one slide from taxel 3 to taxel 1

6.7.2 Sliding

The slip detection analysis described in Section 6.4.3 and 6.4.4 showed that, when slippage occurred on the surface of a sensor, it was impossible to specify the direction or rate of slide, making processing or control more difficult. By designing an array of taxels, the sliding direction may be assessed by analyzing the changes in output of these taxels.

Fig. 6.34 shows plots obtained when a human slid a fingertip along one row of taxels. Smooth changes in taxel outputs were observed, clearly showing the direction of slide (in this case, the fingertip slides from taxel 3 to taxel 1). Their DWT signals also indicate in advance the taxel on which the slippage is acting. Moreover, during the transition between the output of two neighboring taxels, we observed an *exchange* of outputs when the fingertip slid at the borders of these taxels 3& 2 or 2& 1. This was due to the contacting plane between the fingertip and the surface of the sensor both when sliding *out* of this taxel and *into* the adjacent taxel. Thus, the output of one taxel is decreased, while the output of the neighboring taxel is increased. By looking at the magnified graphs of DWT signals at transition phases, we could easily observe the *overlapping* of two signals, indicating two simultaneous sliding actions on two taxels. At high taxel resolution, it is possible to locate not only the place of slippage, but also its direction and relative velocity.

Thus, using deposits of pectinate electrodes, we could assess the contact location, as well as the sliding direction/velocity, on the fabric sensor without dividing it into smaller pieces than when using normal electrodes. Fabric sensors can therefore cover *any* shape or size of a robot body.

6.8 Discussion

6.8.1 Sensor's Sensitivity

Although the sensors are knitted of yarns made from a mixture of polyester and steel fibers, consistent yarn characteristics are not always obtained. We therefore did not measure the *exact* force exerted on the surface of a sensor, except for *static* loading/unloading. Rather, we utilized the *dynamic* response of the sensor to detect slippage of an object while in contact with the sensor. In short, we regarded the output of this sensor as a *phenomenon* rather than a *value*. This sensor can therefore be used, together with other sensing systems such as load cell and tactile array sensors, in a reliable tactile sensing system that can provide static information about force and localization, and dynamic information about slippage and vibration.

During testing, we found that this sensor was not as sensitive to a light touch applied to its surface. Detection of slippage requires a fairly high load, enabling us to observe significant changes in output. Although this value was not high, about 0.6 N, we aim to reduce it, allowing the sensor to respond to a lighter load. Although increasing the amount of steel fiber in the yarn may increase its sensitivity, the current amount, about 30 %, is the maximum we can obtain due to limited fabrication ability. Moreover, as the percentage increases, the yarn becomes conductive and cannot act as a transducer. We therefore intend to change the design of the sensor, particularly the knitting method, to increase its sensitivity. Despite this, we found that the sensor was remarkably sensitive to the touch of a conductive object, such as a human fingertip. We have therefore proposed a new structure, in which the system utilizes a *capacitive* touch sensing technique to detect a very light touch of a human fingertip.

6.8.2 Stretchability for Pectinate Electrode-Based Sensor

Pectinate electrode-based fabric sensors have many advantages, including rapid response, localized measurement, and usability in a large scale area. Nonetheless, pectinate electrodes on the backside of a sensor cannot bear high stretching strain (over 10%), since they may break. This may be troublesome in applications requiring considerable stretching or a curved surface. This disadvantage may be counteracted by depositing copper pectinate electrodes onto a flexible printed circuit (FPC), followed by wrap it and covering it with a fabric sensor. This may enable both high stretch strain of the sensor and localization of pectinate electrodes. The FPC might not totally cover the complicated surface and the contact between electrodes and backside of the sensor may not be as good as the conventional type. Thus, depending on application, we can choose a suitable design to maximize operation of the fabric sensor.

6.8.3 Interactive Actions

In this research, we have focused on detecting two fundamental actions: pushing and sliding. These two actions are considered sufficient for a robot skin at end-effectors while manipulating objects. Nonetheless, when the sensor is used as an interface between humans and robots, these actions are not sufficient. For example, when sensors cover a large area of a robot body, such as its arms, belly, and back, humans interacting with the robot through touching have various types of contact, including pinching, patting, tickling, stroking, squeezing, scratching, and tapping. If a robot attempts to understand human behavior through touching, these actions must be recognized and classified in order for the robot to properly respond. These actions are combinations of pushing and sliding actions with changes of frequency or magnitude, such as stroking, tapping, and patting; while pinching and squeezing are only possible if the body of the robot is soft. Further investigations on distinguishing among multiple states of human touching are needed.

6.9 Conclusion

We have described our attempts to develop a simple but highly sensitive slip sensor, utilizing the tension-sensitive characteristics of electro-conductive yarn with a novel construction of a fabric sensor. In addition, data processing methods were utilized to assess incipient slippage and distinguish among textures. We also introduced a slip indicator for evaluation of slippage characteristic utilizing discrete wavelet transform, and its usage for extraction of features used in texture discrimination. Future work will focus on the introduction of this sensor to robotic hands to detect states of contact (push, slip) during grasping or manipulating objects, and to study human-robot interactions.

7

Slip Perception Using a Tactile Array Sensor

In this chapter, we present the concept of using image processing for tactile sensors as appropriate tools in localizing/recognizing objects in robotic in-hand manipulation tasks, a concept originally derived from our localized displacement phenomenon (LDP) idea. Our approach operates on a moderately high-resolution intensive array of data obtained from a tactile sensor when a robotic gripper grasps an object that is small relative to the size of the fingers. Instead of using tactile data as an array of discrete numbers, we treat the data as *grayscale* images. By working with successive images from the tactile sensor by exploiting image-processing tools, we are able to extract rich information about the contact situation between an object and the gripper. Experimental results show that from the processed data, one can realize the grasped object's position/orientation, contact shape, especially the *stick-slip condition* on the contact surface, which is derived for the first time by this sensor. We also modeled an object-grasping gripper with tactile feedback for various postures of the object, utilizing Beam Bundle Model, and a corresponding experiment setup to validate computed results. The success of this research once again shows the potential of LDP in soft tactile systems, even for the commercialized sensor used in this chapter.

7.1 Introduction

In the field of robotics, imitating human touch is still a challenging task for anthropomorphic robotic hands. Humans can perform dexterous tasks based on not only vision, but also on rich information obtained from the touch mechanism. Even in cases of visual occlusion, a person can easily assess the characteristics of a grasped object, such as friction, roughness, and its position/orientation within the fingers. This is due to cutaneous mechanoreceptors, that include four functionally distinct types of tactile afferent [1]. These afferents have particularly high densities in the fingertips, bringing dynamical events, such as skin deformation, direction and spatial distribution of

A.-V. Ho and S. Hirai, *Mechanics of Localized Slippage in Tactile Sensing*, Springer Tracts in Advanced Robotics 99,
DOI: 10.1007/978-3-319-04123-0_7, © Springer International Publishing Switzerland 2014

contact forces. Recent research on robotics continues to focus on the creation of robotic hand with humanlike sensory systems to perform dexterous tasks, especially a tactile sensing system. However, to imitate all the human tactile afferents would require a complex fusion of several sets of sensors embedded under robotic skin, and there would still remain tremendous difficulties in terms of compactness and effectiveness. Therefore, recent research tends to create multi-modal robotic skin sensors that obtain information ranging from pressure to temperature sensations [20], [56]. There is another trend that attempts to perceive as much information as possible from a sole-modal tactile sensor of pressure distribution, such as texture [72], object formation and recognition [57], [84]. By using advanced technology such as piezoresistive or capacitive array devices, today's tactile sensors have improved sensitivity and higher spatial resolution, but they are still far from human afferents. Nonetheless, tactile information has been used widely in robotic hands in many potential applications, such as object recognition and the assessing of contact states.

Conventional applications using tactile sensors attempt to extract information on force distribution on a contact surface to ascertain when contact occurs or is broken, and the location of contact. There are also numerous studies working on object recognition using machine-learning techniques, incorporating uncertainties in measurements. [57] have modeled tactile sensors using a point-spread function, and made use of tactile images to obtain local surface information during object exploring. [84] also utilized tactile images of objects, taking advantage of "bag-of-features" in vision to propose a recognition method. A tactile sensing system is also utilized in assisting stable grasping in robotics. [68] showed that a tactile sensor could collect contact location information, enhancing the stability of a system in a peg-in-hole task. Contact point location between object and robotic finger was also mentioned in [85] and [86], as an efficient tool to localize a grasped object when vision was occluded, by using image moments. More recently, the authors in [87] and [88] have been employing tactile sensors attached to robotic hands, in companion with machine-learning techniques (Support Vector Machine, Hidden Markov Model), to estimate the stability of a given grasping task. Results promise tactile feedback could carry meaningful information for stable grasping in a *blind grasp*. Most of these approaches used small, coarse-resolution tactile sensors, therefore the obtained data merely brought discrete and insufficient information about the contact status. With increasingly developing technology, it is promising to create a sensor of high resolution. At that time, a different look at tactile sensors will be required, with more advanced and convenient processing methods, to bring rich and reliable information for recognition and control. Moreover, conventional research on tactile sensors only focuses on normal pressure distribution, and pays little consideration to tangential factors, such as the detection of slip action of a grasped object.

Nowadays, most of well-known robotics hand are equipped with tactile array sensors, such as PR2 [90], Barrett hand [88], Gifu-2 hand [91].

Nonetheless, most of them are small-sized, coarse resolution tactile sensor, therefore obtained data merely brought discrete and insufficient information about contact condition. With increasingly developing technology, it is promising to create sensor with high resolution. At that time, it would require a different look of tactile sensor, with more advanced and convenient processing method, to bring rich and reliable information for recognition and control. Moreover, conventional research on tactile sensor only focuses on normal pressure distribution, while pays less consideration on tangential factors, such as detection of slip action of the grasped object. Holweg *et al.* [92] proposed two methods to detect slippage of a rubber-based tactile matrix sensor. The first one employed frequency analysis of movement of the center of force distribution; while the second one utilized fluctuation of normal force thanks to rubber elasticity. Nonetheless, those method were implemented at very low frequency (0.5 Hz), which is neither promising in real time applications, nor timely to fast slip events such as dropping. Recently, Alcazar and Barajas [93] proposed a real-time algorithm utilizing optical flow to estimate slid vectors of grasped object sliding via pressure array sensors. However, this method is only appropriate given the slippage while impossible to detect the incipient slippage. In short, it is necessary to endow an off-the-shelf tactile arrayed sensor with slip/incipient slip perception. This will bring the sensor with multiple functions other than pressure information, so that it would be widely used in the robotic research community.

In this chapter, we show how we exploit image-processing techniques in tactile data information reasoning to bring the benefit of reducing the data processing burden, particularly to the enrichment of information on contact statuses, promising an efficient tool for the implementation of robotic manipulation tasks. Our method consists of treating tactile data as images, based on a good resolution sensing array, conveying multiple modalities of physical contact: pressure, contact shape, and localization. In particular, our innovation is to assess stick-slip condition on a contact surface by tracking micro slips of feature points on a contact surface. We also propose an efficient method to determine whether a grasped object slips, enhancing stable grasping by a robotic hand. For ease of investigation, the grasped object in this chapter is considered to be smaller than the fingertips, which dominates most cases in object manipulation, *i.e.* when the object changes posture its imprint on tactile sensor is observable.

7.2 From Tactile Sensing to Image Processing

7.2.1 Tactile Sensor

In this research, we utilized a Nitta I-SCAN50 tactile sensing system [70] that consists of a grid of tension-sensitive electroconductive ink lines (Fig. 7.1). The working principle is quite straightforward: when no load applies at one

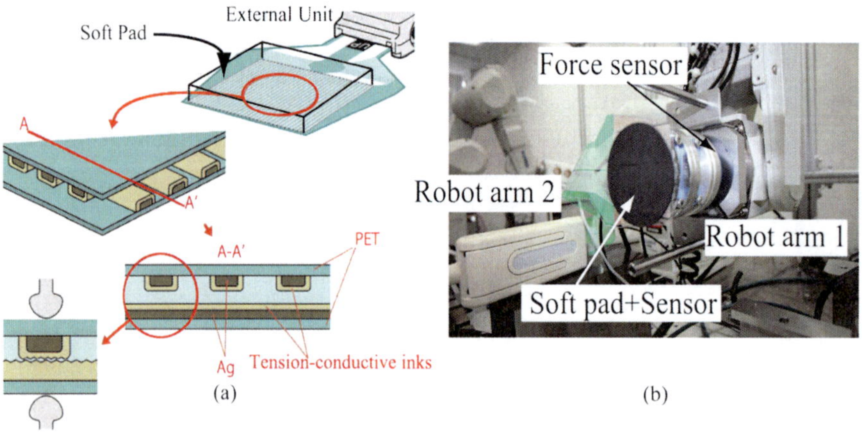

Fig. 7.1 Dual robot arms with tactile sensors at end-effectors to imitate a robotic gripper: (a) I-SCAN tactile sensor with a soft pad. Cross-sectional A-A' shows that each conductive ink covers a silver layer, on a PET (polyethylene terephthalate) sheet. When a load is applied to one node, contact between two inks is formed; the larger the load is, the smaller the value of resistance of the inks. These changes are sensed by an external data-processing unit. (b) Robot arm setup.

intersection, there is light contact between the inks, resulting in a high resistance of the inks; when a force is applied to the sensing sheet, the conductive inks are pushed together to make a strong contact, causing the resistance to drop dramatically. By scanning the intersection nodes, information about pressure distribution can be obtained. This sensor consists of 44×44 *taxels* (tactile elements), measuring 44 mm×44 mm with a 1 mm row/column spacing. A *soft pad* of similar size and 2 mm thick covers the sensing area to form a complete soft tactile fingertip. This tactile system was attached to a robotic finger via a Nitta 6-DOF force/torque sensor. We set up the systems on dual robot arms with an object grasped between their end-effectors, so that they are able to work in cooperation to imitate a robotic parallel gripper.

7.2.2 From Tactile Data to Image Processing

Assume that an object is being grasped by fingertips, resulting in its imprinting the soft pad of a tactile sensor. Depending on the applied load, each taxel has a different 8-bit value, resulting in a 44×44 dimensional array of tactile data as illustrated in Fig. 7.2. Wherever the load applied, the corresponding node will output a non-zero value. In several cases, the sensor will return a positive error (nonzero at no-load state) due to the noise of the measurement and process. We found that the obtained array shown in Fig. 7.2 *looks akin* to a *grayscale image*. This image has a size of 44×44, in which each pixel

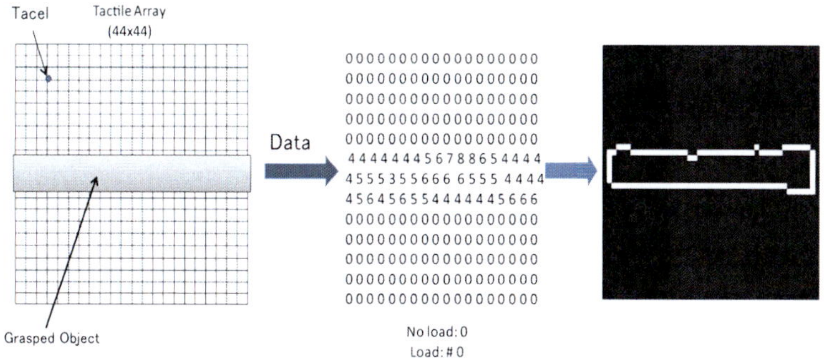

Fig. 7.2 The similarity between tactile data and a grayscale image. A grasped object will leave an imprint on the tactile sensor described by a non-zero value in the output data matrix, and its corresponding illustration on a processed tactile image.

corresponds to a taxel; each pixel has an 8-bit value equal to that of the taxel. In addition, this image has a *background* consisting of zero-valued pixels, and a *foreground* that is formed by nonzero-valued pixels. As a result, a tactile array can be transformed totally into a real grayscale image, known as a *tactile image*. Hence, every action performed on tactile data will be considered as an image-processing operation. In the next section, we perform several examples to see the potential in object manipulation tasks.

The idea of treating tactile data as tactile image, however, is not new if we look back to publications in 1980s such as [94] and [95], or recently such as [57] and [84]. Nonetheless, only static images were employed to retrieve information of normal force distribution, moving images analysis has not been applied to dynamic range. Therefore, with fine resolution tactile array and image processing tool have merited us to have a deeper analysis of tactile image in dynamic application that has not been issued to date, especially slip perception.

7.3 Conventional Applications

7.3.1 Localization

In image-processing techniques, we are able to localize not only the position of the contact, but also its orientation by exploiting the *image moment*, defined as:

$$M_{pq} = \sum_{x} \sum_{y} x^p y^q I(x, y), \tag{7.1}$$

in which $x, y, I(x, y)$ are the coordinates of each taxel in the image coordinates, and its intensity, respectively. Here p is the x-order and q is the

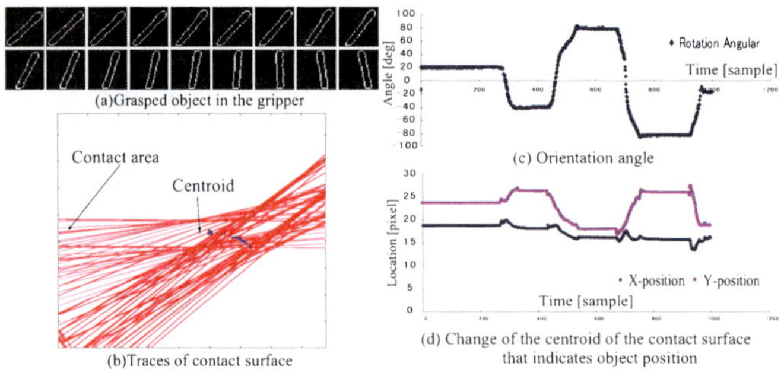

Fig. 7.3 Localization. By using an image-processing tool, the position and localization of an object can be obtained based on calculating moments.

y-order, whereby *order* means the power to which the corresponding component is taken in the above sum [46]. In this case, M_{00} is the contact area; and the centroid of the contact area is computed as follows:

$$\begin{bmatrix} x_0 \\ y_0 \end{bmatrix} = \frac{1}{M_{00}} \begin{bmatrix} M_{10} \\ M_{01} \end{bmatrix}. \tag{7.2}$$

Using this idea of image moment, we also can estimate the orientation of the contact area, through *central moments*:

$$\mu_{pq} = \sum_x \sum_y (x - x_0)^p (y - y_0)^q I(x, y). \tag{7.3}$$

The eigenvectors of the covariance matrix derived by using second-order central moments correspond to the major and minor axes of the image intensity, thus the orientation θ can be extracted from the angle of the eigenvector with the largest eigenvalue by the following equation:

$$\theta = \frac{1}{2} \arctan \left(\frac{2\mu_{11}}{\mu_{20} - \mu_{02}} \right). \tag{7.4}$$

To illustrate this example, an experiment was conducted in which the gripper tightly held a rectangular object, then this object was moved arbitrarily, but while maintaining contact, to imitate an in-hand manipulation task (Figure 7.3(a)). Figure 7.3(b) illustrates the trace of the contact area during the movement of the object in the gripper. We also can obtain the values for the object's orientation angle (Fig. 7.3(c)), and the coordinates of the centroid in the tactile image (Fig. 7.3(d)). Consequently, given the tactile data one can easily localize the grasped object's position and orientation using the above idea of image moments.

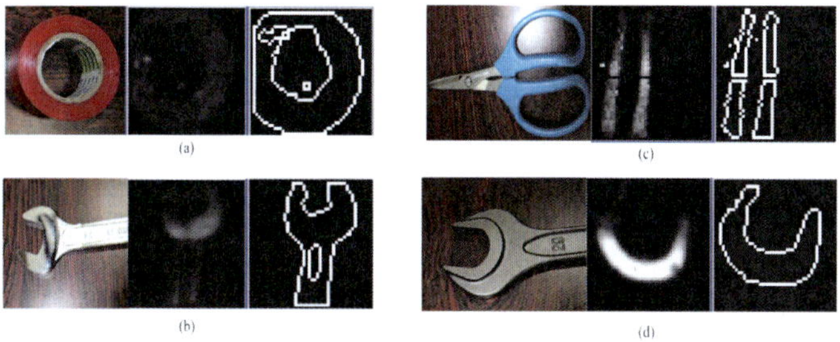

Fig. 7.4 Better look, better recognition. Extracted boundaries of grasped objects bring information about the geometrical shapes of objects entirely in (a) and (b), or partly in (c) and (d).

7.3.2 Contact Shape Recognition

Previous research used discrete information through a specific number of tactile images during an object-exploring task to realize/discriminate objects. Usually, the obtained tactile images were coarse, thus little information about the object could be assessed. Therefore, a complicated teaching/learning method is required to assist the object-recognition process [84]. Naturally, richer, clearer, and more reliable information after each touch would decrease the complication of the learning algorithm, as well as uncertainties, and accelerate the realization process in a real-time application. By exploiting our method with an edge-detection technique, one can expect a better description of the contact area, also a partial shape of the grasped object (Fig. 7.4).

7.4 Slip Perception from Tactile Images

One of the most difficult tasks for tactile sensors is to detect slips occurring during contact with an object. Tactile sensors typically supply information about the normal force distribution of an applied load. Therefore, tangential traction cannot be measured via a change in normal force, resulting in complications in slip detection. One might evaluate the slip through movements of the entire contact surface, such as a change in the centroid's coordinates and/or rotation angle. However, micro slips, which dominate the pre-slide stage of a soft object or non-uniform contact pressure distribution, cannot be attained easily. Our innovation focuses on tracking featured points on the contact area using image processing to detect the incipient/overt slip of the object.

7.4.1 Sub-pixel Slippage

For a tactile array sensor, the underlying information is always the normal force distribution for all types of application. Therefore, it was supposed

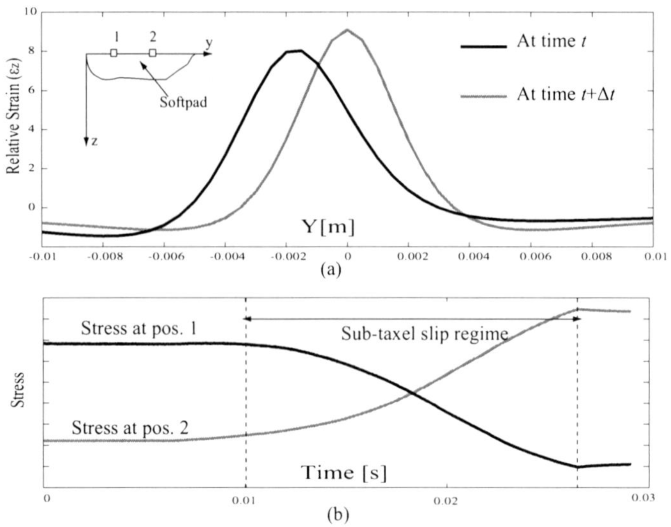

Fig. 7.5 (a) Strain for line load at two locations. (b) Dynamic change of stress of two taxels at position 1 (x, y_t) and 2 $(x, y_{t+\Delta t})$ when the grasped object moves between two positions.

impossible to assess traction on the surface of the sensor. Nonetheless, with a soft pad on top of the sensor, the sensor is expected to be able to "sense" the continuous transition of normal load during a sliding motion by detecting the exchange of normal force information among neighboring taxels. Assume that the soft pad is an homogeneous elastic, and there is a line load applied perpendicularly to the surface of the pad along the z-axis with the contacts are assumed to be infinitely long along the x-axis. Due to the softness, adjacent taxels near the line also suffer deformations, causing non-uniform distribution of plane strain. The relative strain ϵ_z at time t and position y_t along y-axis can be estimated as follows (see [83]):

$$\epsilon_z = \frac{-3P}{2\pi E(z^2 + y_t^2)^2} z(z^2 - y_t^2), \tag{7.5}$$

where P is normal force, and E is Young's modulus. At the time $t + \Delta t$, the line is moved to the new position of $y_{t+\Delta t}$ while keeping the total normal force unchanged. Two distributions of relative strains of the line load at two positions are shown in Fig. 7.5(a). One might see a superposition of two distributions. Therefore, if we consider the dynamic change of stresses obtained from plane strains at the current position $(x, y_{t+\Delta t})$ and the previous position (x, y_t), we can observe the gradual exchange that is illustrated in Fig. 7.5(b). This is referred to as a sub-taxel level of slippage of the grasped object, thanks to the ductility of the soft pad covering the tactile sensor. This suggests that by using any means to detect this exchange, the possibility of slip detection of the grasped object at the sub-taxel level is promising.

7.4.2 Localized Displacement Phenomenon

We were inspired by the localized displacement phenomenon (LDP) presented in earlier chapters. The phenomenon states that during the stick-to-slip transition of a sliding soft fingertip there exists partial movements of contact points, and that overt slippage occurs if and only if all the contact points give way. We also suggested that if we could assess this phenomenon by a sensing system, the slippage of soft objects could be detected in a timely way. We employed a high-speed camera to track the movements of contacting points during the transition paradigm. This, however, cannot be always implemented, especially in robotic hands, due to the complications of camera setup and calibration, as it requires fiducial remarks on the contact surface. Similar experimental attempts to detect the incipient of soft fingertip can be found in [8]. In our application, the sensor is covered by a soft pad, and makes contact with objects having complicated shape and textures, thus the LDP was expected to appear in a grasping task with the occurrence of slippage. As a result, we exploited this phenomenon to propose a way to detect slippage using a tactile array sensor with a soft pad.

7.4.3 Our Approach

With these considerations in mind, we proposed a method to detect the slippage of a grasped object using tactile sensor. Assume that we obtain a set of *feature points* on the imprint of a grasped object in a tactile image. Due to the LDP, there is a number of *slipped* points, in conjunction with *stick* points. Thus, a ratio between the number of slipped points and the number of feature points quantifies the slippage of the grasped object. Based on this ratio, it is possible to judge how and when the slippage occurs. Moreover, as this slippage can be detected at the sub-taxel level, *i.e.*, prior to the actual movement, this is promising in assess movement before it actually happens. In the next section, we report on the implementation of our idea of utilizing an image-processing tool by extracting feature points on the contact imprint, then tracking their movements in a dynamic way.

7.5 Implementation of Slip Detection

7.5.1 Role of Size in Detection of Feature Points

An original tactile image inherits the resolution of the tactile array sensor, thus it is a rather small size (44×44). We attempted to extract feature points of this image by utilizing the corner extraction method. Most corner-finding

Table 7.1 Detected Feature Points over Interpolation Factor

Interpolation Factor	Feature Points
×1	9
×2	26
×4	62
×6	75
×8	82

algorithms work by using an autocorrelation matrix over a predetermined window around each point. A sufficiently large window results in reliable corner detection, which in turn downgrades the number of detected corners in a small image. Consequently, the original tactile image produces an insufficient number of feature points for tracking. To overcome this shortage, we interpolated the original image into a bigger one at factors of 2, 4, 6; then calculated the number of detected feature points. Table 7.1 shows the detected feature points of tactile images when the size varies. One might argue that a bigger image results in more detected feature points. However, we found that the factor exceeded 6, the detector did not perform better. Taking into account the cost of computation, we decided to process on the interpolated tactile image with a factor of 4.

7.5.2 Role of Noise

There are many sources of noise affecting tactile images, such as mechanical noise, analog-to-digital quantization noise, and the different sensitivities of individual taxels. Moreover, the tactile sensor sheet is usually covered by a soft rubber layer that spreads applied force across the surface of the sensor, *i.e.* even when a sole point of contact is made, its adjacent taxels are also activated (see Section 7.4.1). As a result, the obtained data of one taxel is synthesized by the actual load applied to its surface, and the noise caused by the above sources [57]. In this research, we neither attempted to filter out nor smooth the noise, instead we utilized it as an efficient cue to define featured points on the contact surface. In image-processing techniques, good features to track are often *corners* that contain enough information to be chosen from one frame to the next [46]. Corners strongly relate to a change of intensity of neighboring pixels. Therefore, if there was no noise, the intensities of taxels would be coincident across the contact area when a uniform force distribution was applied, causing difficulties in detecting feature points. Consequently, in order to extract sufficient feature points we have to incorporate noises during processing, in conjunction with our proposed slip detection method.

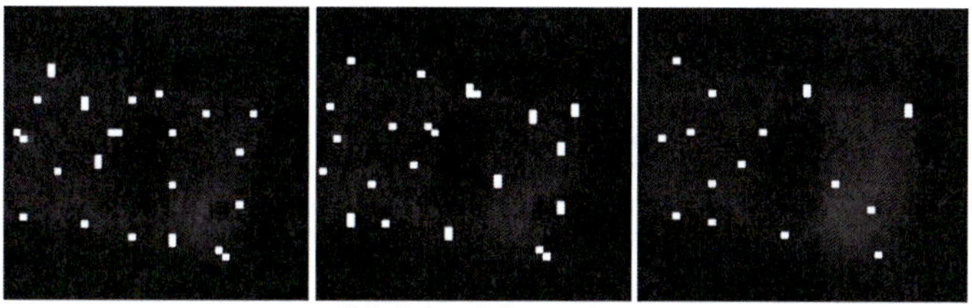

Fig. 7.6 Feature points detected on the imprint of a grasped object on a tactile image at various Gaussian kernels. The number of feature points decreases at a high kernel, *i.e.* better filtering.

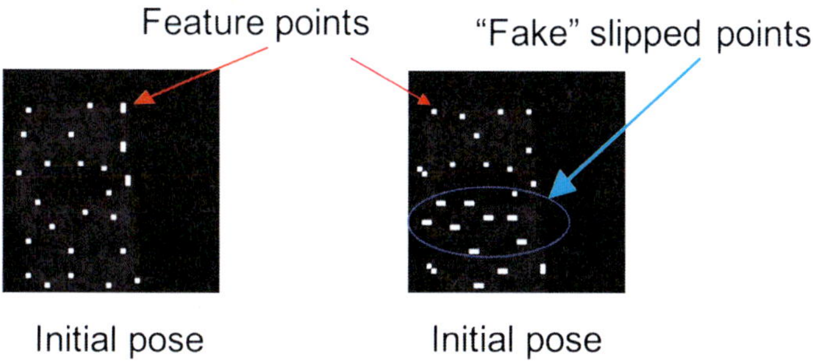

Fig. 7.7 Positive error in tracking movements of feature points. Even during the initial pose, due to noise, some fixed points are detected as 'fake' slipped points.

7.5.3 Slip Detection Method

Assume that a grasped object was moved within the confines of two fingertips of a robotic gripper as mentioned in the above section. The obtained tactile image was convolved using a Gaussian filter at several common kernels, then a Shi and Tomasi corner detection algorithm ([46]) was applied to finally output feature points. Figure 7.6 indicates that the more the image was smoothed, the fewer the number of detected feature points. This strongly supports our idea of incorporating noise in this application. After assessing feature points, the optical flow of these points is extracted using the Pyramid Lucas-Kanade algorithm [46]. However, due to noise, there are fixed feature points detected as slipped points, even at the initial phase (Fig. 7.7). Therefore, a condition for judging a slip point need to be proposed to assure that the algorithm only tracks the moved points. Assuming that a feature point j is detected with coordinates of (x_t, y_t) at time t, and $(x_{t+\Delta t}, y_{t+\Delta t})$ at the next moment $t + \Delta t$. Define the distance between two points:

$$d = \sqrt{(x_t - x_{t+\Delta t})^2 + (y_t - y_{t+\Delta t})^2}. \tag{7.6}$$

The condition for slip/fixed point is expressed as follows:

$$\begin{cases} d > \Psi : Slip \\ d < \Psi : Fixed \end{cases}, \tag{7.7}$$

where Ψ is slip threshold. This can be referred to as the Euclidian distance threshold. This threshold is usually specified empirically at the sub-pixel level, and it depends on the friction resistance at the contact surface, and uncertainties in image processing.

As a result, events of slippage of the grasped object can be detected by analyzing consecutive movements of feature points. To assess analytically slippage of the object, we attempt to introduce a *slip indicator* that represents the quantity of slippage on the contact surface over frames. Assuming that $\{\Omega\}$ is a set of feature points detected at time t, and $\{\alpha\}$ is set of slipped points at time $t + \Delta t$ according to the condition (7.7). Let us define slip indicator, μ, as follows:

$$\mu = \frac{n_\alpha}{n_\Omega}, \tag{7.8}$$

where n_α is the number of slipped points, and n_Ω is the number of detected feature points. We can assess that the slip indicator receives a large value when there are partial slippages on the contact surface, and reaches 1 if an overt slide occurs. However, depending on the movement of the grasped object, the slip indicator is not always necessarily 1. The relative movement of the grasped object is categorized into two main factors: translation and rotation, and both indicate the slippage of the object. Therefore, we attempted to characterize these movements and its influence to the slip indicator in the later part of this chapter.

7.5.4 Translational Slip

In order to characterize the effect of the translational movement of the grasped object on the slip indicator, we conducted an experiment in which the movement of the object was given by a motorized linear stage, and the displacement was recorded by a laser displacement sensor, acting as a ground truth for the experiment. Figure 7.8 shows the response of the slip indicator over the slip movement of the grasped object. One can observe that when the object remains fixed within the gripper, the slip indicator fluctuates around zero. When the object moves, the slip indicator jumps quickly, and keeps a high value around 1 until the object stops moving. In particular, due to localized displacement, the slip indicator responds *prior* to the ground truth changing its value, *i.e.* before the actual slippage of the object. This is assumed to be helpful in judging when the slip occurs in a timely manner. If t_d is the delay time between the onset of the slip and the first moment of the sensor's response (see Fig. 7.10), one might decide the occurrence of the object's slippage during this period. Figure 7.9 shows the variation of t_d when

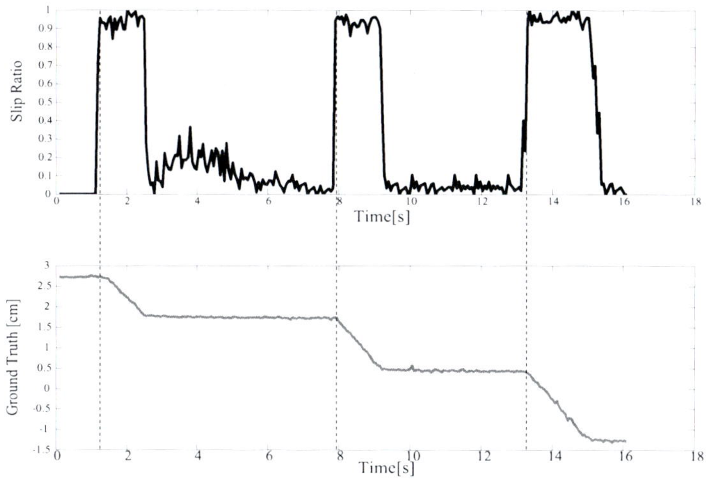

Fig. 7.8 Timely response of the slip-ratio when the grasped object moves

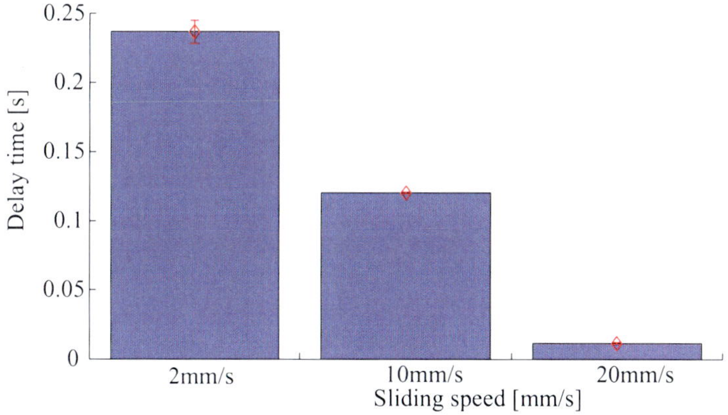

Fig. 7.9 Delay time varies with its standard deviation at different sliding velocities

the sliding speed changes, in an intuitive way, the speedier the sliding motion, the smaller the value of t_d. This method is also sufficient to detect speedy slippage of the grasped object, similar to a dropping event under gravity. Figure 7.10 shows the corresponding response of the slip indicator over the movement of the object, and the sharp rise of the slip indicator during the pre-slide regime indicates the incipient slippage of the grasped object. This immoderation occurs within 20 ms. This is supposed to be sufficient for a sensor responding to a slippage event (see [96]). Therefore, the slip indicator is promising in assessing the incipient slip action of the grasped object, which is crucial to obtaining stable manipulation in robotics research.

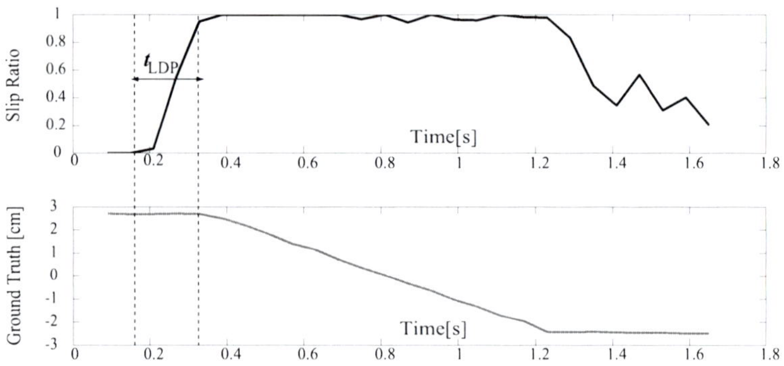

Fig. 7.10 Response of the slip indicator in extremely fast motion

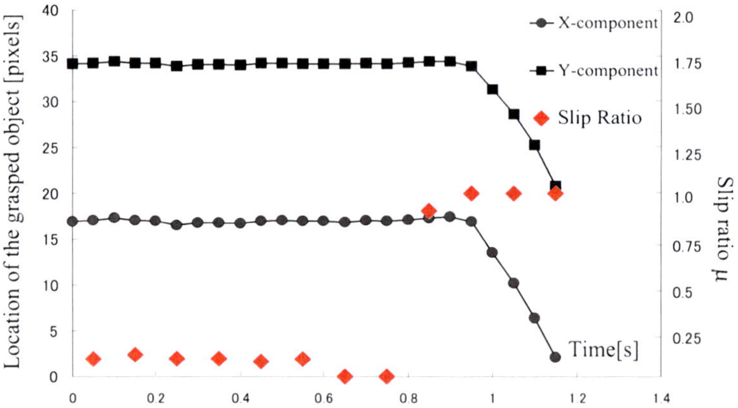

Fig. 7.11 Response of the slip indicator in the case of a fast translational movement of the object. Translational slippage can be observed via the linear descents of the X, Y-components. The slip indicator jumps quickly to the extreme value of 1, and is faster than the centroid.

In conventional use, one might employ the change of a grasped object's imprint location to determine whether the object has moved. We compared the performance of the slip detection of the slip indicator and our proposed method. In the image-processing technique, we are able to localize not only the position of the contact, but also its orientation by exploiting *image moment* M_{pq}, where p is the x-order and q is the y-order indicating the power to which the corresponding component is taken in the above sum [46]. The centroid of the contact area is computed as follows:

$$\begin{bmatrix} X_0 \\ Y_0 \end{bmatrix} = \frac{1}{M_{00}} \begin{bmatrix} M_{10} \\ M_{01} \end{bmatrix}. \tag{7.9}$$

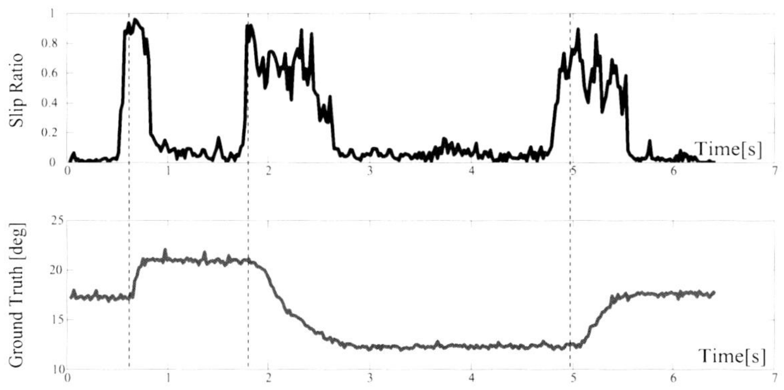

Fig. 7.12 Response of the slip indicator when the grasped object is rotating, given by the ground truth

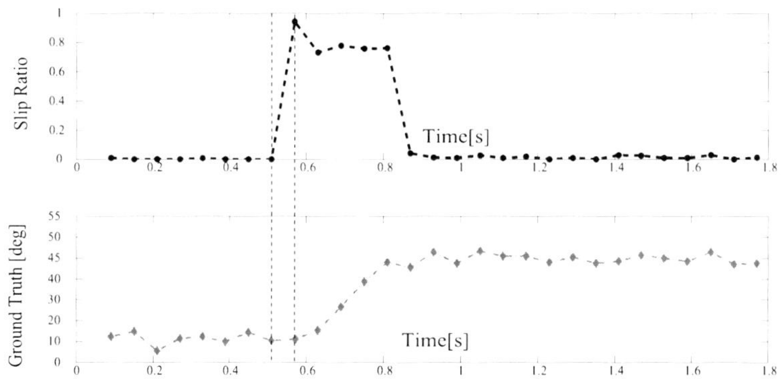

Fig. 7.13 Response of the slip indicator in an extremely fast rotational motion

As a result, using the above technique, the location of the object's imprint can be perceived by specifying the centroid. Figure. 7.11 plots the centroid's X- and Y-components and the response of the slip indicator during one fast translational movement of the grasped object. The centroid responds slower than the slip indicator, in this case approximately 0.2 s. Thus, the slip indicator outperforms the centroid in terms of fast response. Moreover, with a complex-shaped object, change of the centroid is unpredictable, thus it is difficult to judge the onset of slippage. In contrast, the slip indicator is rather robust with respect to the texture and shape of the grasped object.

7.5.5 Rotational Slip

In conjunction with translational movement, the rotation of the grasped object between fingers is typical, especially when the grasped object collides with the outside environment causing a change in the object's posture. Thus, the detection of rotation movement is assumed to be important. Similar to the above section on translational movement, we attempted to create a ground truth for the characterization of the slip indicator for rotational motion. We employed a $\pm 5g$ 3-axis accelerometer, and attached it to the tip of a grasped object. Signals from the accelerometer were transformed into the rotational angle of the grasped object around its rotation axis. The measured angular information is with respect to the coordinates attached to the end effector of a robot finger covered by a tactile sensor, and similar to the actual rotation of the grasped object. Figure 7.12 illustrates the slip indicator when the object's posture is changed under pure rotation as indicated by the ground truth in degrees. Similarly to the translational case, the slip indicator performs well with the onset of slippage caused by the rotation. The delay time t_d also can be seen as an indication for assessing the incipient slip (see above section) even for a fast rotation case as illustrated in Fig. 7.13. In particular, we can see that the slip indicator does not reach the value of 1 even when overt rotation is given, which is different than for translation movement. This is due to the fact that there are always fixed points around the rotation axis of a grasped object, resulting in a smaller number of detected slipped points. Consequently, the slip indicator is maintained at a high value, but less than 1 when rotation occurs. This can be considered a clue to distinguishing translation and rotation movements of the object.

As mentioned, the slip-detection method exploits the appearance of noise to enhance feature detection, thus the effect of noise on the method sometimes counters its reliability. For example, a fast load/unload of grip force might cause mistaken slip detection. We conducted an experiment in which a cylindrical object is held by the gripper and the gripping force is controlled so as to be applied/released fast and suddenly. This event is similar to the collision case or the noise caused by a mechanical system. At that time, the imprint area on the tactile image is widened and shrinks continuously, resulting in feature points being detected as slipped points. The slip indicator also increases corresponding to these events, even when there is no slippage. Figure 7.14 shows plots of slip indicators in a common slip case (dashed line) and in the above case. One can observe that there are sparks caused by the fast load/unload of grip force, but their values are rather small compared to the common case. Therefore, in order to avoid ambiguities, it is necessary to pick a suitable value for the threshold to judge the slippage. Similarly to the translation case, we found that the use of a slip indicator is more timely than the use of change in the contact area's centroid in assessing the onset of the slippage. The slip indicator was also recognized to be robust with respect to the contact area's shape and texture.

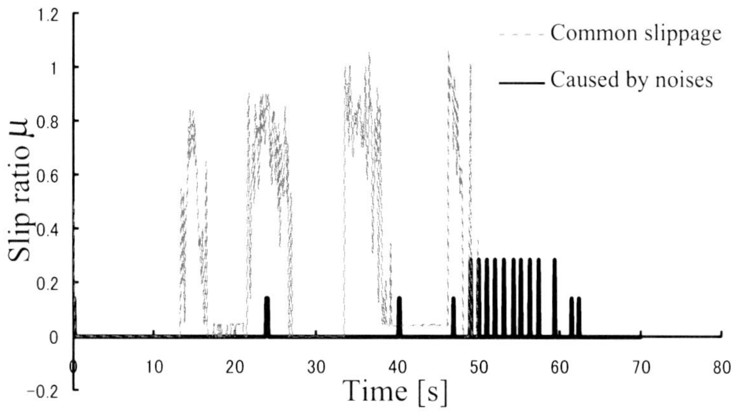

Fig. 7.14 Response of slip indicator for a fast load/unload of the gripping force in which the object does not move (dark solid line). This event creates peaks, causing ambiguity with respect to actual common slip (dashed line). By setting a suitable threshold, we can eliminate this ambiguity.

Consequently, by incorporating noise into tactile images to assess feature points and a slip indicator that is akin to a dynamic friction coefficient, partial and overt slippage of the grasped object can be fully described and detected. The response of the slip indicator is seen to be timely to the onset of the slippage, and fairly distinguishable in terms of an object movement's type (rotation or translation), and robust with respect to a grasped object's texture and shape. The method employed in the implementation of this idea is basically straightforward, and usable by any other researcher. This is a new contribution to endowing pressure sensors with the slip perception that were usually presumed to be lacking in a high enough degree to represent slip traction. The tracking movements of feature points across a contact surface shows promise as a way of calculating shear strain on the contact surface during slip action of the object, however, this requires a lot of work on a robust tracking method. All the above analysis were conducted in real time.

7.6 A Case Study: Force/Torque Estimation Given Tactile Sensing Data

In this section, we estimate the force/torque (F/T) acting on robotic fingertips at various postures for a grasped object given information from a tactile sensor during in-hand manipulation, utilizing BBM. A model is proposed for the tactile sensor, and the estimated force/torque wrench is compared to real ones from a 6-DOF F/T sensor.

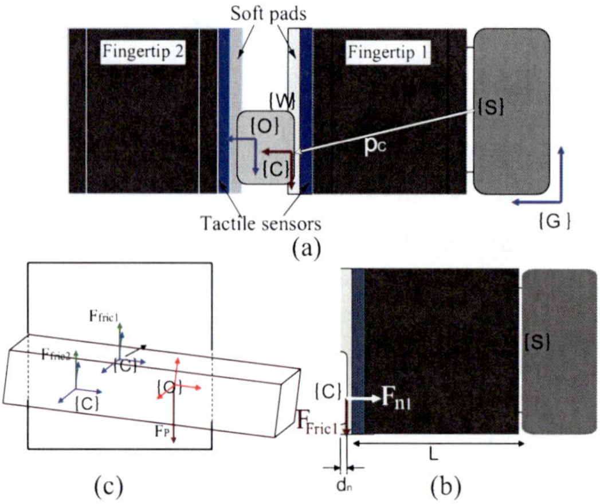

Fig. 7.15 Geometrical analysis: (a) Object grasped with a gripper. (b) Separation of one finger for analyzing the replaced reactive forces acting at the origin of the $\{C\}$-coordinates. (c) Localization of the grasped object within the gripper, and the external forces acting on it.

7.6.1 Estimation Model

Let a known object be grasped by a gripper with an arbitrary posture as illustrated in Fig. 7.15(a). One can see in Fig. 7.15 that $\{O\}, \{C\}, \{W\}, \{S\}, \{G\}$ are coordinate systems (or frames) of the object, contact location, wrist, sensor location, and a global one,respectively. We use feedback from a tactile sensor to estimate the wrench$[\mathbf{F}, \mathbf{M}]^T$ acting on the position of the frame $\{S\}$ (location of F/T sensor). Figure 7.15(c) shows the relative position of the grasped object in the gripper. In the coordinate system $\{W\}$, localization of the object is specified by the position and orientation of the frame $\{C\}$ with respect to $\{W\}$, particularly x_c, y_c, z_c, θ_c. While (x_c, y_c, θ_c) can be easily obtained through the localization ability of the tactile sensor mentioned previous sections; the contact depth of the object over the soft pad $z_c = d_n$ can be estimated by the relative position between two fingers of the gripper. The external forces acting on the object are contact force, including normal and tangential components, and the gravity force.

For ease of analysis, we detach virtually the grasped object from the fingertips, and replace it with reactive force acting on the contact surface. This reactive force is split into two components: a normal component and a tangential component. This force is considered to be placed at the centroid of the contact surface, which is the placement of the $\{C\}$-coordinate system (see Fig. 7.15(b)). By estimating these components, we are able to assess the F/T acting on the fingertip.

To calculate normal force component acting on the contact surface, we use the Beam Bundle Model (BBM), in which the soft pad is virtually divided

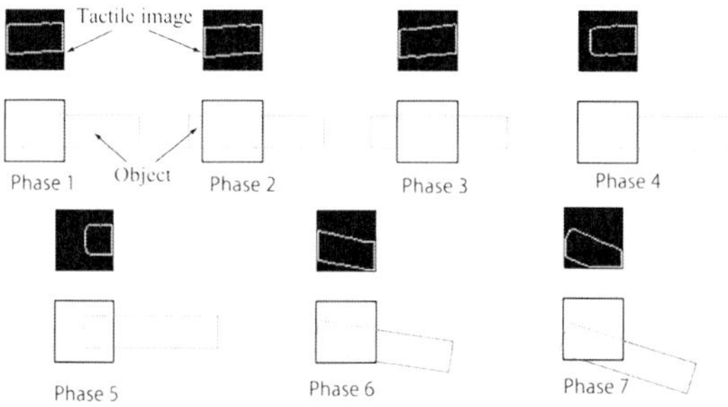

Fig. 7.16 Postures of the grasped object through phases. In this application, the localization of the object's contact surface is estimated by tactile images.

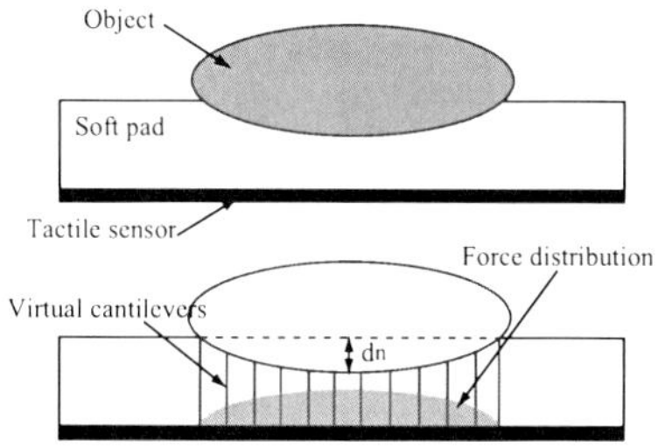

Fig. 7.17 Proposed model for assessing normal force distribution. The soft pad is considered to consist of virtual beams that are compressible and bendable. When the load is applied, the deformations at the contact surface are transformed to those of the beams, resulting in a normal force distribution depending on how each beam deforms.

into an infinite number of elastic beams. Normal force distribution can be estimated by calculating the reactive force acting on the deformed beams (Fig. 7.17). In this chapter the contact surface is flat, thus the beams have the same contact depth d_n (Fig. 7.15(b)), which simplifies the calculation of the normal force as stated below:

$$|\mathbf{F}_{n_1}| = \int\int_S \frac{Ed_s}{l} d_n = \frac{E}{l} d_n \int\int_S d_s = \frac{ES}{l} d_n, \tag{7.10}$$

where E, l are the Young's modulus and the thickness of the soft pad, respectively, and S is the contact area that is calculated as zero-order moment M_{00} of the tactile image mentioned in Section 7.3.1. As a result, the normal force component in the frame $\{C\}$ is specified as $\mathbf{F}^C_{n_1} = [0\ 0\ F_{n_1}]^t$. In this estimation, we ignored the superposition of stresses when the object is applied to the soft pad of the sensor as mentioned in [83], which results in a larger value for the normal force than the estimated one. Nonetheless, for the purpose of preliminary estimation, this is acceptable.

There are two components of friction force acting on the two contact surfaces of fingers. Due to the symmetry, it is sufficient to calculate one, and the other is obtained similarly. In the coordinate system $\{O\}$, we have the following equations:

$$\begin{cases} \mathbf{F}^O_{fric_1} + \mathbf{F}^O_{fric_2} + \mathbf{F}^O_P = \mathbf{0} \\ |\mathbf{F}^O_{fric_1}| = |\mathbf{F}^O_{fric_2}| = |\mathbf{P}/2|. \end{cases} \tag{7.11}$$

As a result, friction force in the frame $\{C\}$ is given as:

$$\mathbf{F}^C_{fric_1} = \mathbf{T}^C_O(-\frac{1}{2}\mathbf{F}^O_P) = -\frac{1}{2}\mathbf{T}^C_O\mathbf{F}^O_P, \tag{7.12}$$

where $\mathbf{T}^C_O$ is a homogeneous transform matrix from frame $\{O\}$ to frame $\{C\}$. Consequently, the forces acting on the frame $\{S\}$ are estimated as follows:

$$\mathbf{F}^S = \mathbf{T}^S_C\mathbf{F}^C_{contact} = \mathbf{T}^S_C(\mathbf{F}^C_{n_1} + \mathbf{F}^C_{fric_1}), \tag{7.13}$$

where $\mathbf{T}^S_C$ is the homogeneous transform matrix from frame $\{C\}$ to $\{S\}$.

To estimate the moments $\mathbf{M}$ on the $\{S\}$-coordinate system, we simply introduce a vector $\mathbf{p}^S_C = [p_x\ p_y\ p_z]^t$ from $\{S\}$ to $\{C\}$ so that: $\mathbf{M} = \mathbf{p}^S_C \times \mathbf{F}^S$. By using a skew symmetric matrix $\mathbf{P}^S_C$ we can easily obtain the following relation: $\mathbf{M} = \mathbf{p}^S_C \times \mathbf{F}^S = \mathbf{P}^S_C\mathbf{F}^S$. Because this analysis is static, friction torque acting on the contact surface can be eliminated. As a result, for each posture of the grasped object and data from the tactile sensor, wrench acting on the fingertip can be obtained as follows:

$$\begin{bmatrix} \mathbf{F} \\ \mathbf{M} \end{bmatrix} = \begin{bmatrix} \mathbf{T}^S_C\mathbf{F}^C_n + \mathbf{T}^S_C\mathbf{F}^C_{fric} \\ \mathbf{P}^S_C\mathbf{T}^S_C\mathbf{F}^C_n + \mathbf{P}^S_C\mathbf{T}^S_C\mathbf{F}^C_{fric} \end{bmatrix}. \tag{7.14}$$

7.6.2 Experiment Results

In this experiment, we change the posture of the grasped object in sequence as illustrated in Fig. 7.16. The obtained wrenchfrom the F/T sensor (see Fig. 7.1(b)) is then compared with the calculated results from the estimation model in Eq. 7.14. Some representative results are plotted in Fig. 7.18. One can see that there is a similarity between the estimated wrench and the experimental wrench. There are slight differences stemming from the uncertainties in the experimental setup, data acquisition, and estimation model. As a result, this case study shows that by using tactile data only and the proposed estimation model, we can obtain the force/torque acting on the fingertip preliminarily.

7.7 Discussions

7.7.1 Threshold

As mentioned in Section 7.5.3, to be able to distinguish "fake" slipped points, it is necessary to employ an Euclidian distance threshold Ψ. The selection of Ψ

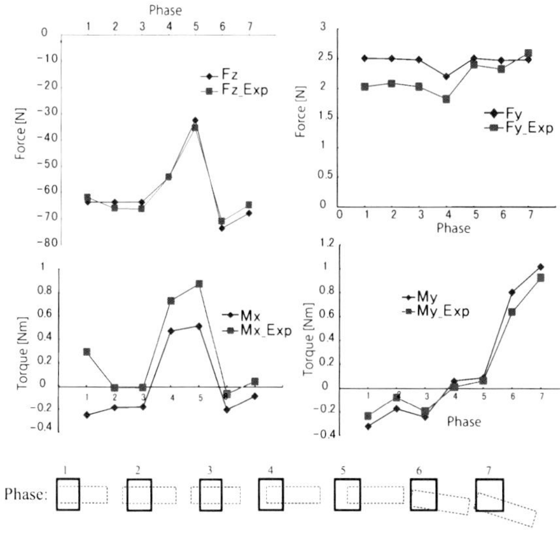

Fig. 7.18 Comparison between estimated wrench (dark blue dots) and experimental ones (gray dots)

Table 7.2 SIFT and SURF Performance

Detector	Feature Points	Computation Cost [s]
SIFT	30	3.8
SURF	17	2.2

depends on how fast we want to detect the occurrence of sub-taxel slippage as discussed in Section 7.4.1. Usually it is chosen as a value less than the distance between two adjacent taxels. Moreover, the exchange time in Section 7.4.1 depends on the mechanical characteristics of the soft pad covering the tactile sensor, such as stiffness and homogeneousness; for example, too stiff a material would result in a small value of Ψ, and vice versa. As a result, Ψ must be tuned depending on the configuration of the tactile sensor, such as resolution and mechanical characteristics. We conducted an analysis of the effect of Ψ on slip perception of the algorithm. The change in slip indicator is seen as depending greatly on the threshold Ψ, as illustrated in Fig. 7.19; in the way that a small value of Ψ might cause *positive* error (see Fig. 7.19(a)), *i.e.* slip is detected when it actually does not occur; whereas a large value of Ψ easily results in *negative* error (see Fig. 7.19(b)), *i.e.* slip is not judged even when it occurs.

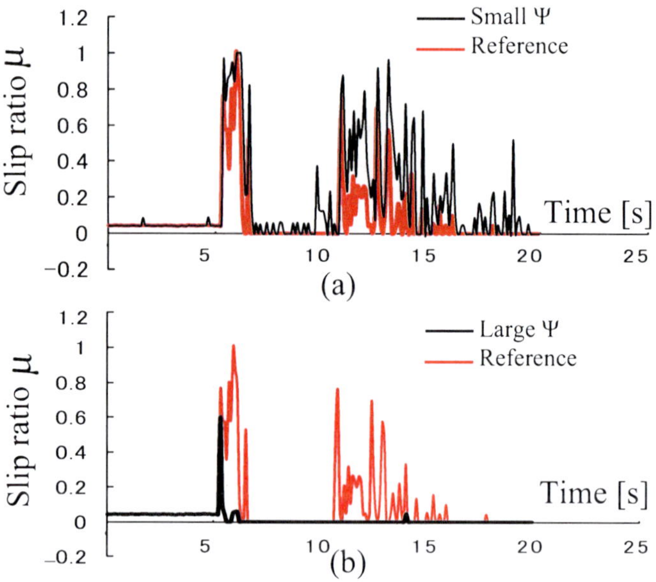

Fig. 7.19 Effect of Ψ on slip indicator: (a) When Ψ is small. (b) When Ψ is large.

Fig. 7.20 SURF-based matched feature points of tactile images at different interpolation factors: (a) Factor of 2, (b) Factor of 4

7.7.2 Robust Tracking

In this chapter, the grasped object is smaller than the size of the tactile sensor, thus detected feature points are extracted easily by a corner-based detection algorithm. The number of feature points depends on the shape and texture of the object's contacted area. In several cases when the object is relatively large compared to the size of the sensor, or has little texture, the existing algorithm does not perform well. Thus, it is necessary to detect inner feature points that are invariant to scale, rotation, and so on. We chose SURF (Speed-UP Robust Feature, [98]), and SIFT (Scale-Invariant Feature Transform, [97]) to investigate their performances on tactile images (with an interpolation factor of 4). The number of feature points detected using SIFT

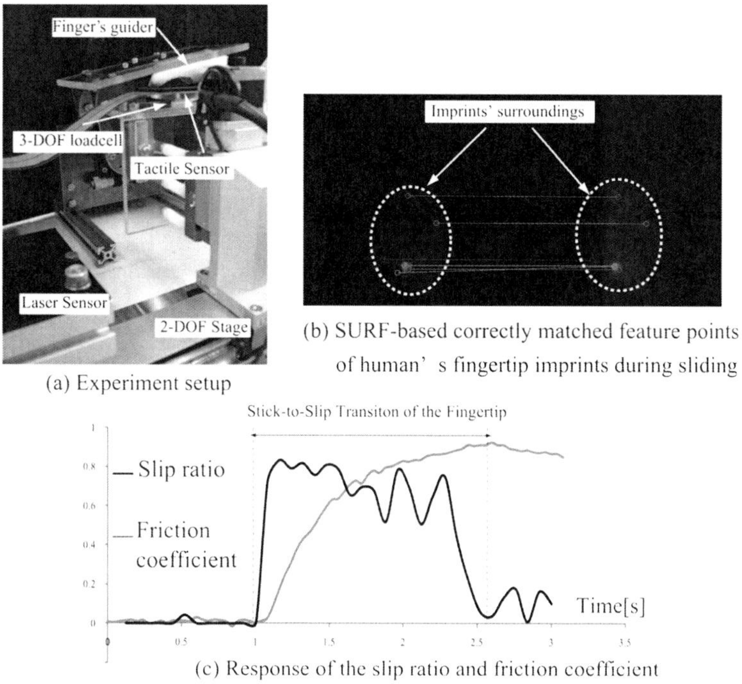

(a) Experiment setup

(b) SURF-based correctly matched feature points of human' s fingertip imprints during sliding

(c) Response of the slip ratio and friction coefficient

Fig. 7.21 Experiment for characterization of human fingertip sliding using a tactile array sensor

was larger than that using SURF, however, the cost is more expensive in term of calculation time (see Table 7.2). As we are going to conduct a robust tracking algorithm in a real-time application, SURF was chosen as the cost is low. Figure 7.20 shows the performance of SURF in the detection of inner points of the contact area. With an application requiring high performance in terms of feature detection and tracking, a robust tracking using SURF feature points can be exploited to speed up the calculation time in a real-time implementation.

7.7.3 Human Finger Slip Detection

A tactile sensor with slip perception is not only usable in robotic manipulation, but also efficient in the detection of the slippage of a human finger in a human-robot or haptic interface, especially for a pointing device. Assessing the slip of the human fingertip by using a camera to capture skin stretch on the contact area has been employed in several projects (*e.g.* see [50]). While this method merits are a clear image of the contact area, resulting in reliable tracking; it typically causes complications in term of the experiment setup and portability of the implementing device. A human fingertip when about to slide is also subject to the LDP, thus we attempted to detect the incipient slip of a human fingertip utilizing the above system and algorithm. We conducted an experiment in which a human finger was fixed on a frame, and a plate covered by the tactile sensor was controlled by a motorized 2-DOF stage to control contact and slide of the fingertip (see Fig. 7.21(a)). As the imprint of the human fingertip is rather simple, we exploited SURF to detect inner feature points on a tactile image interpolated by the factor of 4. Figure 7.21(b) shows the matched feature points on tactile images from which one might see that the detected number is rather small, and the successfully matched rate is about 65 % (7/11). The preliminary result is shown in Fig. 7.21(c), which plots the slip indicator during the stick-to-slip transition of the fingertip, and shows that the slip indicator reacts properly to the onset of the slippage indicated by a change in the dynamic friction coefficient calculated by the load cell's signals, and during the stick-to-slip transition of the fingertip. In short, using this setup, it is promising to create a pointing device to realize the incipient slip motion of human finger for a haptic device. In the future, we will implement SURF by eliminating most incorrectly matched points to obtain a more robust tracking, and conduct experiments with many subjects in real-time mode.

7.8 Conclusions

We have presented our approaches to the enhancement of a tactile array sensor utilizing an image-processing tool for in-hand manipulation. Usually, a tactile sensor has limited use in static applications such as determination of normal stress distribution, contact location, partial recognition of object, yet dynamic changes. By using our method, not only can conventional tasks be performed easily, but also dynamic tasks can be assessed in a timely way. Slippage of the grasped object is, for the first time in a tactile array sensor, fully detected by the proposal of slip indicator utilizing image processing tool. This research is originally derived from the LDP idea.

In the future, we will build a library for processing general tactile sensors with the functions mentioned in this thesis. It is envisaged that this library could be used for any tactile sensor for a specific application. At the same time, multiple applications will be carried out. First, object recognition utilizing supervised learning methods will be implemented to equip the sensing system with recognition ability. Second, a fusion of a tactile sensor and a F/T sensor will be employed to estimate reliable information of a grasped object during in-hand manipulation, exploiting probabilistic filters such as Kalman filter or particle filter. Finally, we will embed this sensing system into robotic hands to enhance the performance during manipulation with proposed tactics.

8

Concluding Remarks

8.1 Conclusion

Our objective in this book was to propose a dynamic model of a sliding soft fingertip to understand in detail how and when slip occurs on a contact surface, from which we can assess the slip perception of soft tactile systems. Our model is simplified to be implemented with less computation cost, dynamically, and be re-creatable by other researchers. This model is implemented under the scenario of a unilateral sliding motion on a flat rigid contact surface of a homogeneous soft fingertip.

Our general approach was to use a Beam Bundle Model (BBM) to model a soft fingertip using the Finite Element Method (FEM), which allowed us not only to represent diverse deformation of the fingertip during a sliding motion, but also to track micro movements on the contact surface. In doing so, we chose to focus mainly on the stick-to-slip transition, *i.e.* the pre-slide phase, then discover the Localized Displacement Phenomenon (LDP) that dominates this transition. By studying this phenomenon, which shows that during the transition phase there exist both slipped and stick points on the contact surface, we could understand slip perception, then propose slip-detection methods for several soft tactile systems.

We assumed that the interior of the soft fingertip consisted of a bundle of elastic beams, which allowed us to simulate deformation of the fingertip as well as to reduce remarkably computation time. In addition, we introduced two methods to constrain the beam bundle according to 2-D and 3-D models of the soft fingertip. In the 2-D case, interactions of beams were represented by a linkage element using a Voigt model that constrain the movements of the beams' contact ends to unilateral movements. In the 3-D case, we modeled the interaction on the contact surface using FE analysis to utilize shear stress relation on the contact patch. Moreover, we introduced Coulomb's friction law at each contact point, as we wanted to characterize friction on the contact patch of the soft fingertip for the purpose of control. As a result, we were

A.-V. Ho and S. Hirai, *Mechanics of Localized Slippage in Tactile Sensing*, Springer Tracts in Advanced Robotics 99,
DOI: 10.1007/978-3-319-04123-0_8, © Springer International Publishing Switzerland 2014

able to simulate dynamic stick-to-slip transition, and answer the question as to how and when slip occurs on a soft fingertip.

The second part of this book focused on the fabrication of soft tactile systems, and endowing them with slip and texture perception based on the LDP idea. These systems, which spread from self-developed devices to off-the-shelf products, have the unique abilities of slip detection after being modeled and characterized using BBM and LDP. In the design of a soft tactile fingertip with an embedded micro force/moment sensor (MFMS) chip, the system exacts the force/moment exerted on it, but is too slow to detect slippage. By assigning the sensor at the origin of the Great Beam, and implementing a sliding trial simulation, we were able to equip the system with slippage detection that was generated by localized displacement on the contact surface. The second system was designed based on the LDP by taking all the slippage of piles on the contact patch during stick-slip transition into account to generate slippage detection. We also implemented various data-processing methods and machine-learning methods to endow this sensor with slip and texture perception, and we found out that discrete wavelet transform (DWT) method is highly suitable for this sensor. The commercial tactile array sensor was finally investigated during my internship at the Mitsubishi Electrical Corporation. We were inspired by utilizing LDP into finding slip-detection ability for this sensor. As a result, we used image-processing techniques to "see" localized displacements on the contact surface, and proposed a slip ratio to judge whether slip occurs. We are confident to conclude that, by utilizing BBM and LDP, we could characterize most typical soft tactile systems, and endow them with the slip-detection ability that is crucial for robotic manipulation.

8.2 Summary of Contributions

The contributions of this book are as follows:

We presented the first formulation of the BBM to model a sliding soft fingertip with friction. In addition, we also successfully represented theoretically, for the first time, the LDP during the stick-slip transition of a fingertip. Using BBM with constraint methods such as FEM helped to simulate dynamically the 3-D deformation of a soft fingertip with less computation cost compared to commercial FE software such as ANSYS or ABAQUS. With a cylindrical soft fingertip we could implement simulation in real time, while with a hemispherical fingertip it only took several minutes to conduct a simulation trial with a standard desktop computer. This model has been used elsewhere to simulate sliding action of other system, such as a tire's slippage on road.

We then extended the use of BBM and LDP into assessing slip perceptions of soft tactile systems. Thanks to the suggestion prompted by the LDP, we were able to create a simple yet efficient tactile system to detect contact states, slip detection, and texture discrimination. One can expect to exploit

BBM and LDP to create one's own tactile systems endowed with multiple perceptions.

8.3 Future Work

8.3.1 BBM with Friction Torque and Rolling Action

As mentioned in Section 3.6.2, it is necessary to introduce friction torque into the sliding soft fingertip model. We were able to estimate friction torque around the normal axis caused by unsymmetrical movements of contacting points, but we have not yet verified it experimentally. In the future, we will conduct a simulation of the soft fingertip when sliding on a complicated path, and characterize the response of friction force and friction torque with verification by an experimental setup.

We have not conducted a simulation with rolling effects, although it is likely practical for this model. When a rolling action occurs, some new points will make contact with the surface, while some contacting points will no longer stick. Moreover, the structures of virtual beams also change when the contact angle varies during a rolling motion. Although it may increase computational cost, it is not a problematic issue if we could accelerate a simulation. As a result, by continuously observing possible contact points during a rolling motion, it would be possible to introduce a rolling effect into this model

8.3.2 A BBM Platform

In this book, we only conducted simulations and experiments on different shaped soft fingertips: cylindrical and hemispherical ones. We hope to create a platform utilizing BBM so that it can be used to model a wide range of soft fingertips with various outer shapes, such as the human fingertip. We have used magnetic resonant images and A snake algorithm to specify the exact shape of fingertips, as well as the locations of bone and other tissues. Based on this information, the stiffness of each beam corresponding to bone, soft tissue, etc., can be determined. Physical connections between beams of different stiffness are made to assure the flawlessness of the model. Contact surfaces are meshed, corresponding to the characteristics of human skin. Thus, by extending our proposed model, we can assess the simulation of the sliding of a human finger. We also wish to accelerate simulation trials by utilizing power of parallel computation, using CUDA in GPU-equipped computers. It is very promising that we can implement a simulation in real time in the near future.

8.3.3 Improvement of the Slip Fabric Sensor and Application

During testing, we found that this sensor was not so sensitive to a light touch applied to its surface. Detection of slippage requires a fairly high load, enabling us to observe significant changes in output. Although this value was not high, about 0.6 N, we aim to reduce it, allowing the sensor to respond to a light load. Although increasing the amount of steel fiber in the yarn may be expected to increase its sensitivity, the current amount, about 30 %, is the maximum we can obtain due to limited fabrication ability. Moreover, as the percentage increases, the yarn becomes conductive and cannot act as a transducer. We therefore intend to change the design of the sensor, particularly the knitting method, to increase its sensitivity. We are currently fabricating this new prototype and we will report on it in the near future.

8.3.4 Tactile Array Sensor under Uncertainties and Intelligence

To conduct full simulations of robotic hand manipulation tasks, it is necessary to simulate responses from tactile sensor. One of the most difficult points is that real tactile array sensors inherently contain noises at taxels, and calibration for taxels remains challenging. Moreover, tactile sensors are conventionally covered by soft pads, which causes trouble for modeling. In the future, we will attempt to a model tactile sensor, tackling these issues as follows:

1. Exploiting the BBM paradigm to modeling deformation of a soft pad. In doing so, given contacting positions, normal force distribution should be estimated properly regardless how of complex the contact surface.
2. Introducing uncertainty to the output of each taxel by smoothing it with an added probabilistic Gaussian convolution filter. This function will represent the interaction between surrounding points and the loaded point that is conventional in a tactile array sensor.

We also aim to fuse the tactile sensor, force-torque sensor, and encoder of a robotic hand. These sensing systems inherently exist in the hands, in contrast to distant vision system, however, yet popular in applications due to instability, limit in scope, especially tactile sensors. The fusion system may present a probabilistic framework for estimating various contact formations of a grasped object, as well as its location in hand or space, under uncertainties. Prior to that, we must select informative measurement features for measurement equations. Ultimately, we want to perform dexterous manipulation tasks by exploiting the fusion of sensors.

A

Continuous Modeling of 2D Elastic Deformation

In this chapter, we will formulate two-dimensional (2D) / three-dimensional (3D) elastic deformation. Stress and strain consist of not only normal components but also shear components. Finite element approximation is applied to formulate potential and kinetic energies of a deformable object to derive the equations of deformation.

A.1 Two-Dimensional Finite Element Modeling

A.1.1 Piecewise Linear Approximation

Let us approximate function $u(\xi, \eta)$ by a piecewise linear function. Let S be the domain of the function. Cover region S by a set of triangles, as illustrated in Fig. A.1. Let P_i be a nodal point of a triangle. Let $[\,\xi_i,\ \eta_i\,]^{\mathrm{T}}$ be coordinates of point P_i. Let $\triangle \mathrm{P}_i \mathrm{P}_j \mathrm{P}_k$ be a triangle consisting of nodal points P_i, P_j, and P_k.

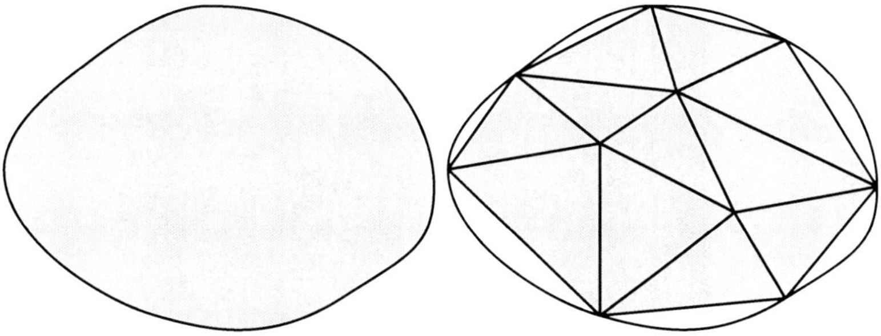

(a) region (b) cover by triangles

Fig. A.1 Cover of two-dimensional region by triangles

We introduce the signed area of a triangle. The area is positive if the triangular loop is counter clockwise while is negative if the loop is clockwise. The signed area of triangle $\triangle \mathrm{OP}_i\mathrm{P}_j$ is then described as follows:

$$\triangle \mathrm{OP}_i\mathrm{P}_j = \frac{1}{2}\begin{vmatrix} \xi_i & \xi_j \\ \eta_i & \eta_j \end{vmatrix} = \frac{1}{2}(\xi_i\eta_j - \eta_i\xi_j).$$

Then, the signed distance of triangle $\triangle \mathrm{P}_i\mathrm{P}_j\mathrm{P}_k$ is given by

$$\begin{aligned}\triangle \mathrm{P}_i\mathrm{P}_j\mathrm{P}_k &= \triangle \mathrm{P}_i\mathrm{P}_j\mathrm{O} + \triangle \mathrm{OP}_j\mathrm{P}_k + \triangle \mathrm{P}_i\mathrm{OP}_k \\ &= \triangle \mathrm{OP}_i\mathrm{P}_j + \triangle \mathrm{OP}_j\mathrm{P}_k + \triangle \mathrm{OP}_k\mathrm{P}_i.\end{aligned}$$

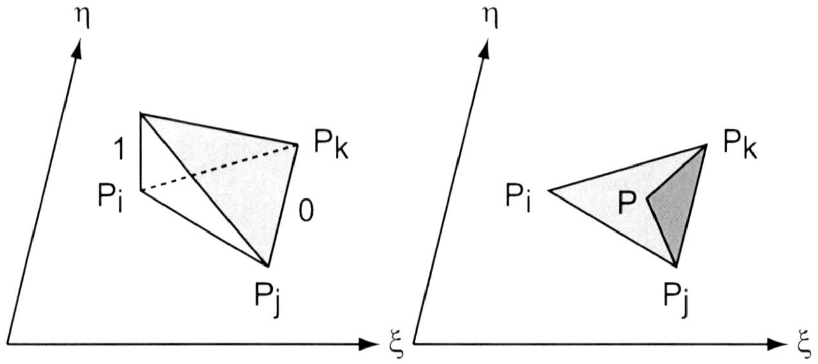

(a) value (b) ratio of areas

Fig. A.2 Function $N_{i,j,k}(\xi,\eta)$

Define two-dimensional shape function $N_{i,j,k}(\xi,\eta)$ on triangle $\triangle \mathrm{P}_i\mathrm{P}_j\mathrm{P}_k$. Assume that the area of the triangle is positive. Function $N_{i,j,k}(\xi,\eta)$ takes 1 at point P_i and takes 0 at point P_j and P_k. Let P be an arbitrary point within the triangle. Let $[\xi,\eta]^{\mathrm{T}}$ be coordinates of point P. As illustrated in Fig. A.2, ratio between areas of $\triangle \mathrm{PP}_j\mathrm{P}_k$ and $\triangle \mathrm{P}_i\mathrm{P}_j\mathrm{P}_k$ determines the function:

$$N_{i,j,k}(\xi,\eta) = \frac{\triangle \mathrm{PP}_j\mathrm{P}_k}{\triangle \mathrm{P}_i\mathrm{P}_j\mathrm{P}_k} = \frac{\triangle \mathrm{OPP}_j + \triangle \mathrm{OP}_j\mathrm{P}_k + \triangle \mathrm{OP}_k\mathrm{P}}{\triangle \mathrm{P}_i\mathrm{P}_j\mathrm{P}_k}$$

Consequently,

$$N_{i,j,k}(\xi,\eta) = \frac{\xi(\eta_j - \eta_k) - (\xi_j - \xi_k)\eta + (\xi_j\eta_k - \xi_k\eta_j)}{2\triangle \mathrm{P}_i\mathrm{P}_j\mathrm{P}_k}. \quad \text{(A.1)}$$

Note that function $N_{i,j,k}(\xi,\eta)$ vanishes outside $\triangle \mathrm{P}_i\mathrm{P}_j\mathrm{P}_k$. Any function $u(\xi,\eta)$ can be linearly approximated as follows inside $\triangle \mathrm{P}_i\mathrm{P}_j\mathrm{P}_k$:

$$u(\xi,\eta) = u_i N_{i,j,k}(\xi,\eta) + u_j N_{j,k,i}(\xi,\eta) + u_k N_{k,i,j}(\xi,\eta), \quad \text{(A.2)}$$

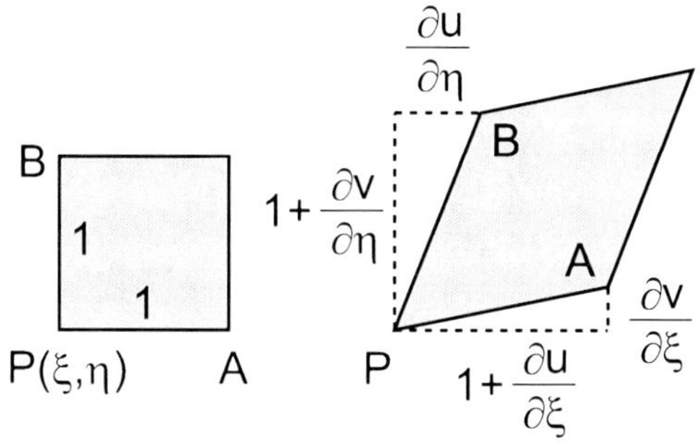

(a) initial shape (b) deformed shape

Fig. A.3 Deformation of small square region

where $u_i = u(\xi_i, \eta_i)$, $u_j = u(\xi_j, \eta_j)$, and $u_k = u(\xi_k, \eta_k)$. Partial derivatives of the function are given by

$$\frac{\partial N_{i,j,k}}{\partial \xi} = \frac{\eta_j - \eta_k}{2\triangle \mathrm{P}_i \mathrm{P}_j \mathrm{P}_k}, \qquad \frac{\partial N_{i,j,k}}{\partial \eta} = \frac{-(\xi_j - \xi_k)}{2\triangle \mathrm{P}_i \mathrm{P}_j \mathrm{P}_k}. \tag{A.3}$$

A.1.2 Two-Dimensional Elastic Deformation

In this section, we will formulate 2D elastic deformation and apply finite element method to compute the deformation numerically. Let P be an arbitrary point on an elastic object. Let $\mathrm{C} - \xi\eta$ be a frame attached on the object. Let (ξ, η) be the coordinates of point P in the object frame. Assume that spatial coordinates of point P are described as (ξ, η) at the initial shape of the object. Deformation of the object can change the spatial coordinate of point P. Let $u(\xi, \eta)$ denotes the displacement of point $\mathrm{P}(\xi, \eta)$ along ξ-axis and $v(\xi, \eta)$ denotes the displacement along η-axis. Vector $\mathbf{u} = [\, u, \, v \,]^{\mathrm{T}}$ is referred to as a *displacement vector*. Note that vector $\mathbf{u}$ depends on ξ and η.

Let us describe the deformation of an object using partial derivatives $u_\xi = \partial u / \partial \xi$, $u_\eta = \partial u / \partial \eta$, $v_\xi = \partial v / \partial \xi$, and $v_\eta = \partial v / \partial \eta$. Assume that a square of unit length at point $\mathrm{P}(\xi, \eta)$ deforms to a parallelogram, as illustrated in Fig. A.3. Two edges of the parallelogram are given as follows:

$$\overrightarrow{\mathrm{PA}} = \begin{bmatrix} 1 + \dfrac{\partial u}{\partial \xi} \\ \dfrac{\partial u}{\partial \eta} \end{bmatrix}, \qquad \overrightarrow{\mathrm{PB}} = \begin{bmatrix} \dfrac{\partial v}{\partial \xi} \\ 1 + \dfrac{\partial v}{\partial \eta} \end{bmatrix}.$$

Note that translational motion of an object does not affect the partial derivatives since any translational motion yields constant displacements over the object region. In other words, we can eliminate translational motion from displacement vectors by computing their partial derivatives. Rotational motion affects the partial derivatives since any rotational motion yields non-constant

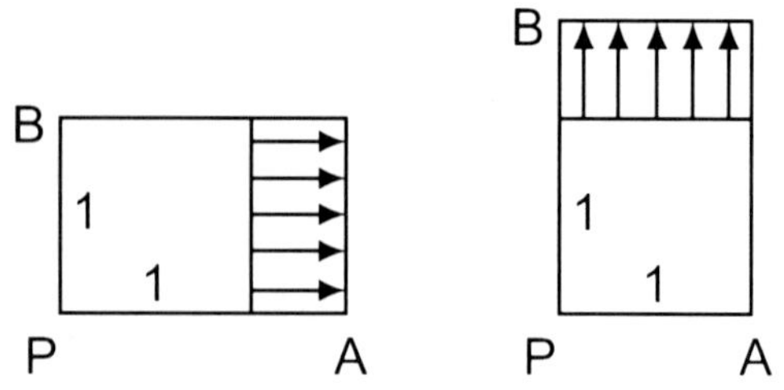

(a) extension along ξ-axis (b) extension along η-axis

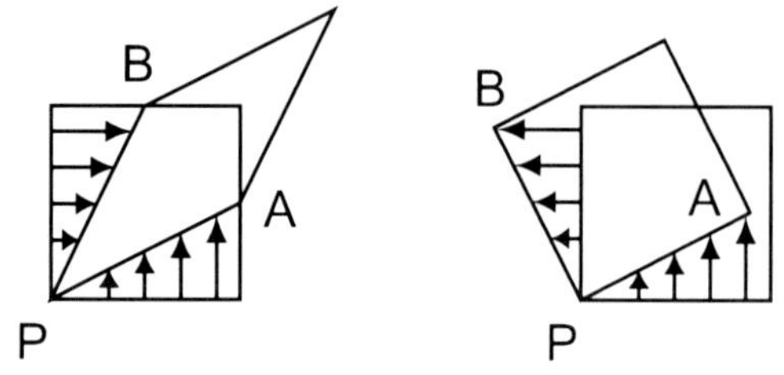

(c) shear deformation (d) rotational motion

Fig. A.4 Deformation and rotation

displacements. In addition, deformation of a 2D object is classified into (a) extension along ξ-axis, (b) extension along η-axis, and (c) shear, as shown in Fig. A.4. Comparing Figs. A.3 and A.4, we have

$$\frac{\partial u}{\partial \xi} = \text{extension along } \xi\text{-axis},$$
$$\frac{\partial v}{\partial \eta} = \text{extension along } \eta\text{-axis},$$
$$\frac{\partial u}{\partial \eta} = \text{shear} - \text{rotation},$$
$$\frac{\partial v}{\partial \xi} = \text{shear} + \text{rotation}.$$

Let $\varepsilon_{\xi\xi}$ and $\varepsilon_{\eta\eta}$ be normal strain components along ξ- and η-axes at point P and $\varepsilon_{\xi\eta}$ be shear strain at point P, we have

$$\varepsilon_{\xi\xi} = \frac{\partial u}{\partial \xi}, \qquad \varepsilon_{\eta\eta} = \frac{\partial v}{\partial \eta}, \qquad 2\varepsilon_{\xi\eta} = \frac{\partial u}{\partial \eta} + \frac{\partial v}{\partial \xi}, \tag{A.4}$$

which are referred to as *Cauchy strain components*, or simply *strain components*. Let us define

$$\varepsilon \triangleq \begin{bmatrix} \varepsilon_{\xi\xi} \\ \varepsilon_{\eta\eta} \\ 2\varepsilon_{\xi\eta} \end{bmatrix}, \tag{A.5}$$

which is referred to as a *pseudo strain vector* or simply *strain vector*.

Let $\sigma_{\xi\xi}$ and $\sigma_{\eta\eta}$ are normal stress components along ξ- and η-axes at point P and $\sigma_{\xi\eta}$ describe a shear stress component at point P. Let us describe these components in a vector form as follows:

$$\sigma \triangleq \begin{bmatrix} \sigma_{\xi\xi} \\ \sigma_{\eta\eta} \\ \sigma_{\xi\eta} \end{bmatrix}. \tag{A.6}$$

Linear elastic material satisfies the following equation:

$$\sigma = D\varepsilon \tag{A.7}$$

where 3×3 matrix D is referred to as an *elasticity matrix*. Note that elasticity matrix D is symmetric. Elasticity matrix of an isotropic material can be described as follows:

$$D = \begin{bmatrix} \lambda + 2\mu & \lambda & 0 \\ \lambda & \lambda + 2\mu & 0 \\ 0 & 0 & \mu \end{bmatrix}. \tag{A.8}$$

where λ and μ denote Lamé's constants. The elasticity matrix can be decomposed into two as follows:

$$D = \lambda I_\lambda + \mu I_\mu \tag{A.9}$$

where

$$I_\lambda = \begin{bmatrix} 1\,1\,0 \\ 1\,1\,0 \\ 0\,0\,0 \end{bmatrix}, \quad I_\mu = \begin{bmatrix} 2\,0\,0 \\ 0\,2\,0 \\ 0\,0\,1 \end{bmatrix}.$$

Matrices I_λ and I_μ originate from the isotropy of the object deformation. Elastic deformation can be characterized by two constants: λ and μ, which determine normal elasticity and shear elasticity.

Lamé's constants are described by Young's modulus E and Poisson's ratio ν as follows:

$$\lambda = \frac{\nu E}{(1+\nu)(1-2\nu)}, \quad \mu = \frac{E}{2(1+\nu)}. \tag{A.10}$$

Additionally, constant μ is equal to shear elasticity modulus G.

A.1.3 Computing 2D Elastic Energy

This section formulates the strain energy of a 2D elastic object. Let S be a region of a 2D elastic object. Assuming that the object is composed of linear elastic material, stain energy of the object is formulated as follows:

$$U = \int_S \frac{1}{2}\sigma^{\mathrm{T}}\varepsilon\, h\, \mathrm{d}S = \int_S \frac{1}{2}\varepsilon^{\mathrm{T}} D\varepsilon\, h\, \mathrm{d}S, \tag{A.11}$$

where h denotes the constant thickness of the object. Partitioning region S into a set of triangles, strain energy is described as

$$U = \sum_{\triangle \mathrm{P}_i \mathrm{P}_j \mathrm{P}_k} U_{i,j,k}$$

where

$$U_{i,j,k} = \int_{\triangle \mathrm{P}_i \mathrm{P}_j \mathrm{P}_k} \frac{1}{2} \varepsilon^{\mathrm{T}} D \varepsilon \, h \, \mathrm{d}S. \tag{A.12}$$

In region $\triangle \mathrm{P}_i \mathrm{P}_j \mathrm{P}_k$, displacement $\mathbf{u}$ can be approximated by a linear combination of nodal displacements $\mathbf{u}_i$, $\mathbf{u}_j$, and $\mathbf{u}_k$ as follows:

$$\mathbf{u} = N_{i,j,k} \mathbf{u}_i + N_{j,k,i} \mathbf{u}_j + N_{k,i,j} \mathbf{u}_k. \tag{A.13}$$

Let $\mathbf{u}_{i,j,k}$ be a collective vector consisting of displacements at P_i, P_j, and P_k:

$$\mathbf{u}_{i,j,k} = \begin{bmatrix} \mathbf{u}_i \\ \mathbf{u}_j \\ \mathbf{u}_k \end{bmatrix}. \tag{A.14}$$

In addition, we will introduce the following collective vectors:

$$\gamma_u \triangleq \begin{bmatrix} u_i \\ u_j \\ u_k \end{bmatrix}, \quad \gamma_v \triangleq \begin{bmatrix} v_i \\ v_j \\ v_k \end{bmatrix}, \quad \gamma \triangleq \begin{bmatrix} \gamma_u \\ \gamma_v \end{bmatrix}. \tag{A.15}$$

Let us calculate strain components inside $\triangle = \triangle \mathrm{P}_i \mathrm{P}_j \mathrm{P}_k$. Substituting Eq. A.13 into Eq. A.4 and applying Eq. A.3, we have

$$\begin{aligned} u_\xi &= \mathbf{a}^{\mathrm{T}} \gamma_u, & u_\eta &= \mathbf{b}^{\mathrm{T}} \gamma_u, \\ v_\xi &= \mathbf{a}^{\mathrm{T}} \gamma_v, & v_\eta &= \mathbf{b}^{\mathrm{T}} \gamma_v, \end{aligned} \tag{A.16}$$

where

$$\mathbf{a} = \frac{1}{2\triangle} \begin{bmatrix} \eta_j - \eta_k \\ \eta_k - \eta_i \\ \eta_i - \eta_j \end{bmatrix}, \quad \mathbf{b} = \frac{-1}{2\triangle} \begin{bmatrix} \xi_j - \xi_k \\ \xi_k - \xi_i \\ \xi_i - \xi_j \end{bmatrix}. \tag{A.17}$$

Then, normal and shear strain components are described as follows:

$$\varepsilon_{\xi\xi} = \mathbf{a}^{\mathrm{T}} \varepsilon_{\xi\xi} = \mathbf{a}^{\mathrm{T}} \gamma_u, \quad \mathbf{b}^{\mathrm{T}} \gamma_v, \quad 2\varepsilon_{\xi\eta} = \mathbf{b}^{\mathrm{T}} \gamma_u + \mathbf{a}^{\mathrm{T}} \gamma_v. \tag{A.18}$$

Recalling Eq. A.9, we find that density of potential energy of an isotropic linear elastic material is described as

$$\begin{aligned} \frac{1}{2} \varepsilon^{\mathrm{T}} D \varepsilon &= \frac{1}{2} \varepsilon^{\mathrm{T}} (\lambda I_\lambda + \mu I_\mu) \varepsilon \\ &= \frac{1}{2} \lambda (\varepsilon_{\xi\xi} + \varepsilon_{\eta\eta})^2 + \frac{1}{2} \mu \left\{ 2\varepsilon_{\xi\xi}^2 + 2\varepsilon_{\eta\eta}^2 + (2\varepsilon_{\xi\eta})^2 \right\}. \end{aligned}$$

Integrating the above density over triangular element $\triangle$ yields the potential energy stored in the element.

Let us calculate the potential energy stored in triangular element $\triangle$. Let us introduce the following notations:

$$\xi_{ij} \stackrel{\triangle}{=} \xi_i - \xi_j, \quad \eta_{ij} \stackrel{\triangle}{=} \eta_i - \eta_j.$$

Let us introduce the following integrals:

$$G_\lambda \stackrel{\triangle}{=} \frac{1}{2}\int_\triangle (\varepsilon_{\xi\xi} + \varepsilon_{\eta\eta})^2\, h\, \mathrm{d}S,$$
$$G_\mu \stackrel{\triangle}{=} \frac{1}{2}\int_\triangle \left\{2\varepsilon_{\xi\xi}^2 + 2\varepsilon_{\eta\eta}^2 + (2\varepsilon_{\xi\eta})^2\right\} h\, \mathrm{d}S.$$

Computing integral G_λ yields

$$G_\lambda = \frac{1}{2}\gamma^{\mathrm{T}} L \gamma \tag{A.19}$$

where L is a 6×6 matrix given by

$$L = \begin{bmatrix} L_{aa} & L_{ab} \\ L_{ba} & L_{bb} \end{bmatrix},$$

and

$$L_{aa} = \int_\triangle \mathbf{a}\mathbf{a}^{\mathrm{T}} h\, \mathrm{d}S, \quad L_{bb} = \int_\triangle \mathbf{b}\mathbf{b}^{\mathrm{T}} h\, \mathrm{d}S, \quad L_{ab} = L_{ba}^{\mathrm{T}} = \int_\triangle \mathbf{a}\mathbf{b}^{\mathrm{T}} h\, \mathrm{d}S.$$

Calculating the above integrals, we have

$$L_{aa} = \mathbf{a}\mathbf{a}^{\mathrm{T}} \triangle h = \frac{h}{4\triangle}\begin{bmatrix} \eta_{jk}^2 & \eta_{jk}\eta_{ki} & \eta_{jk}\eta_{ij} \\ \eta_{ki}\eta_{jk} & \eta_{ki}^2 & \eta_{ki}\eta_{ij} \\ \eta_{ij}\eta_{jk} & \eta_{ij}\eta_{ki} & \eta_{ij}^2 \end{bmatrix},$$
$$L_{bb} = \mathbf{b}\mathbf{b}^{\mathrm{T}} \triangle h = \frac{h}{4\triangle}\begin{bmatrix} \xi_{jk}^2 & \xi_{jk}\xi_{ki} & \xi_{jk}\xi_{ij} \\ \xi_{ki}\xi_{jk} & \xi_{ki}^2 & \xi_{ki}\xi_{ij} \\ \xi_{ij}\xi_{jk} & \xi_{ij}\xi_{ki} & \xi_{ij}^2 \end{bmatrix},$$
$$L_{ab} = \mathbf{a}\mathbf{b}^{\mathrm{T}} \triangle h = \frac{h}{4\triangle}\begin{bmatrix} -\eta_{jk}\xi_{jk} & -\eta_{jk}\xi_{ki} & -\eta_{jk}\xi_{ij} \\ -\eta_{ki}\xi_{jk} & -\eta_{ki}\xi_{ki} & -\eta_{ki}\xi_{ij} \\ -\eta_{ij}\xi_{jk} & -\eta_{ij}\xi_{ki} & -\eta_{ij}\xi_{ij} \end{bmatrix}.$$

Similarly, we have

$$G_\mu = \frac{1}{2}\gamma^{\mathrm{T}} M \gamma, \tag{A.20}$$

where M is a 6×6 matrix given by

$$M = \begin{bmatrix} M_{aa} & M_{ab} \\ M_{ba} & M_{bb} \end{bmatrix},$$

and

$$M_{aa} = 2L_{aa} + L_{bb}, \quad M_{bb} = 2L_{bb} + L_{aa}, \quad M_{ab} = L_{ba}, \quad M_{ba} = L_{ab}.$$

Consequently, potential energy stored in element $\triangle \mathrm{P}_i\mathrm{P}_j\mathrm{P}_k$ is described as

$$U_{i,j,k} = \lambda G_\lambda + \mu G_\mu. \tag{A.21}$$

Note that material property is characterized by Lamé's constants, λ and μ, whereas integrals G_λ and G_μ depend on geometry of the element alone.

Potential energy $U_{i,j,k}$ is denoted with respect to γ. Let us describe the energy with respect to $\mathbf{u}_{i,j,k}$. Introduce permutation matrix P that converts vector $\mathbf{u}_{i,j,k}$ into γ:

$$\gamma = P\,\mathbf{u}_{i,j,k}, \tag{A.22}$$

that is,

$$\begin{bmatrix} u_i \\ u_j \\ u_k \\ \hline v_i \\ v_j \\ v_k \end{bmatrix} = \left[\begin{array}{cc|cc|cc} 1 & & & & & \\ & & 1 & & & \\ & & & & 1 & \\ \hline & 1 & & & & \\ & & & 1 & & \\ & & & & & 1 \end{array}\right] \begin{bmatrix} u_i \\ v_i \\ \hline u_j \\ v_j \\ \hline u_k \\ v_k \end{bmatrix}. \tag{A.23}$$

Integrals G_λ and G_μ are then rewritten as

$$G_\lambda = \frac{1}{2}(P\mathbf{u}_{i,j,k})^{\mathrm{T}} L (P\mathbf{u}_{i,j,k}) = \frac{1}{2}\mathbf{u}_{i,j,k} J_\lambda^{i,j,k} \mathbf{u}_{i,j,k},$$
$$G_\mu = \frac{1}{2}(P\mathbf{u}_{i,j,k})^{\mathrm{T}} M (P\mathbf{u}_{i,j,k}) = \frac{1}{2}\mathbf{u}_{i,j,k} J_\mu^{i,j,k} \mathbf{u}_{i,j,k},$$

where

$$J_\lambda^{i,j,k} = P^{\mathrm{T}} L P, \tag{A.24}$$
$$J_\mu^{i,j,k} = P^{\mathrm{T}} M P. \tag{A.25}$$

As a result, potential energy $U_{i,j,k}$ is described as

$$U_{i,j,k} = \frac{1}{2}\mathbf{u}_{i,j,k}^{\mathrm{T}} K_{i,j,k} \mathbf{u}_{i,j,k}, \tag{A.26}$$

where

$$K_{i,j,k} = \lambda J_\lambda^{i,j,k} + \mu J_\mu^{i,j,k}. \tag{A.27}$$

Matrices $J_\lambda^{i,j,k}$ and $J_\mu^{i,j,k}$ are referred to as *partial connection matrices*. Partial connection matrices are geometric; they include no physical parameters.

Summing up all strain energies at individual triangles, we find that total strain energy is formulated as follows:

$$U = \frac{1}{2}\mathbf{u}_{\mathrm{N}}^{\mathrm{T}} K \mathbf{u}_{\mathrm{N}}, \tag{A.28}$$

where

$$K = \lambda J_\lambda + \mu J_\mu \tag{A.29}$$

is referred to as a *stiffness matrix*. Matrices J_λ and J_μ are referred to as *connection matrices*. Connection matrices J_λ and J_μ are geometric as well. The $(0,0)$-th block of matrix $J_\lambda^{i,j,k}$ contributes to the (i,i)-th block of connection matrix J_λ, the $(0,1)$-th block of $J_\lambda^{i,j,k}$ contributes to the (i,j)-th block of the connection matrix, and so on. Summing up all contributions yields connection matrix J_λ. Similarly, the $(0,0)$-th block of matrix $J_\mu^{i,j,k}$ contributes to the (i,i)-th block of connection matrix J_μ, the $(0,1)$-th block of $J_\mu^{i,j,k}$ contributes to the (i,j)-th block of the connection matrix, and so on. Summing up all contributions yields connection matrix J_μ.

Calculation of partial connection matrices

Let us demonstrate the calculation of partial connection matrices of $\triangle \mathrm{P}_i\mathrm{P}_j\mathrm{P}_k$, where $(\xi_i, \eta_i) = (0,0)$, $(\xi_j, \eta_j) = (l, 0)$, and $(\xi_k, \eta_k) = (0, l)$. Since

$$\mathbf{a} = \frac{1}{l^2}\begin{bmatrix} -l \\ l \\ 0 \end{bmatrix} = \frac{1}{l}\begin{bmatrix} -1 \\ 1 \\ 0 \end{bmatrix}, \quad \mathbf{b} = \frac{-1}{l^2}\begin{bmatrix} l \\ 0 \\ -l \end{bmatrix} = \frac{1}{l}\begin{bmatrix} -1 \\ 0 \\ 1 \end{bmatrix},$$

we have

$$L = \left[\begin{array}{c|c} L_{aa} & L_{ab} \\ \hline L_{ba} & L_{bb} \end{array}\right] = \frac{h}{2}\left[\begin{array}{ccc|ccc} 1 & -1 & 0 & 1 & 0 & -1 \\ -1 & 1 & 0 & -1 & 0 & 1 \\ 0 & 0 & 0 & 0 & 0 & 0 \\ \hline 1 & -1 & 0 & 1 & 0 & -1 \\ 0 & 0 & 0 & 0 & 0 & 0 \\ -1 & 1 & 0 & -1 & 0 & 1 \end{array}\right],$$

$$M = \left[\begin{array}{c|c} M_{aa} & M_{ab} \\ \hline M_{ba} & M_{bb} \end{array}\right] = \frac{h}{2}\left[\begin{array}{ccc|ccc} 3 & -2 & -1 & 1 & -1 & 0 \\ -2 & 2 & 0 & 0 & 0 & 0 \\ 0 & 0 & 1 & -1 & 1 & 0 \\ \hline 1 & 0 & -1 & 3 & -1 & -2 \\ -1 & 0 & 1 & -1 & 1 & 0 \\ 0 & 0 & 0 & -2 & 0 & 2 \end{array}\right].$$

Consequently,

$$J_\lambda^{i,j,k} = \frac{h}{2}\left[\begin{array}{cc|cc|cc} 1 & 1 & -1 & 0 & 0 & -1 \\ 1 & 1 & -1 & 0 & 0 & -1 \\ \hline -1 & -1 & 1 & 0 & 0 & 1 \\ 0 & 0 & 0 & 0 & 0 & 0 \\ \hline 0 & 0 & 0 & 0 & 0 & 0 \\ -1 & -1 & 1 & 0 & 0 & 1 \end{array}\right],$$

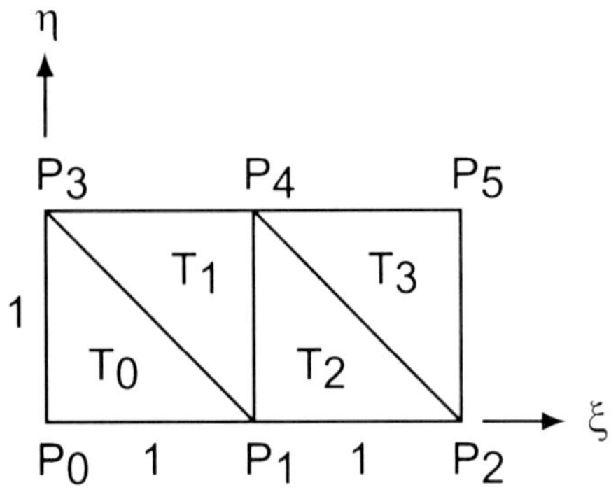

Fig. A.5 Example of 2D elastic object

$$J_{\mu}^{i,j,k} = \frac{h}{2}\left[\begin{array}{rr|rr|rr} 3 & 1 & -2 & -1 & -1 & 0 \\ 1 & 3 & 0 & -1 & -1 & -2 \\ \hline -2 & 0 & 2 & 0 & 0 & 0 \\ -1 & -1 & 0 & 1 & 1 & 0 \\ \hline -1 & -1 & 0 & 1 & 1 & 0 \\ 0 & -2 & 0 & 0 & 0 & 2 \end{array}\right].$$

Note that the partial connection matrices are independent of l.

Example of Connection Matrix Construction

Let us demonstrate the construction of connection matrices J_λ and J_μ by taking a simple example illustrated in Fig. A.5. A square is divided into 4 triangles, T_0 through T_3. Nodal points P_0 through P_5 describe the deformation of the square. Assume that the thickness h is constantly equal to 2.

Computing partial connection matrices corresponding to triangle $T_0 = \triangle P_0P_1P_3$, we have

$$J_{\lambda}^{0,1,3} = \left[\begin{array}{rr|rr|rr} 1 & 1 & -1 & 0 & 0 & -1 \\ 1 & 1 & -1 & 0 & 0 & -1 \\ \hline -1 & -1 & 1 & 0 & 0 & 1 \\ 0 & 0 & 0 & 0 & 0 & 0 \\ \hline 0 & 0 & 0 & 0 & 0 & 0 \\ -1 & -1 & 1 & 0 & 0 & 1 \end{array}\right],$$

$$J_{\mu}^{0,1,3} = \left[\begin{array}{rr|rr|rr} 3 & 1 & -2 & -1 & -1 & 0 \\ 1 & 3 & 0 & -1 & -1 & -2 \\ \hline -2 & 0 & 2 & 0 & 0 & 0 \\ -1 & -1 & 0 & 1 & 1 & 0 \\ \hline -1 & -1 & 0 & 1 & 1 & 0 \\ 0 & -2 & 0 & 0 & 0 & 2 \end{array}\right].$$

Reflecting triangle $\mathrm{T}_0 = \triangle \mathrm{P}_0\mathrm{P}_1\mathrm{P}_3$ with respect to coordinate origin and translating the reflected triangle by 1 and 1 along ξ- and η-axes coincides with triangle $\mathrm{T}_1 = \triangle \mathrm{P}_4\mathrm{P}_3\mathrm{P}_1$. Thus, we find $J_\lambda^{4,3,1} = J_\lambda^{0,1,3}$ and $J_\mu^{4,3,1} = J_\mu^{0,1,3}$Similarly, we find $J_\lambda^{1,2,4} = J_\lambda^{5,4,2} = J_\lambda^{0,1,3}$ and $J_\mu^{1,2,4} = J_\mu^{5,4,2} = J_\mu^{0,1,3}$

Let us compute connection matrices in an iterative manner. First, let J_λ and J_μ be zero matrices. Let us incorporate the contribution of $\mathrm{T}_0 = \triangle \mathrm{P}_0\mathrm{P}_1\mathrm{P}_3$. For example, the $(0,0)$-th block of matrix $J_\lambda^{0,1,3}$ contributes to the $(0,0)$-th block of matrix J_λ and $(0,2)$-th block of matrix $J_\lambda^{0,1,3}$ contributes to the $(0,3)$-th block of matrix J_λ. Adding the contribution of matrix $J_\lambda^{0,1,3}$ to connection matrix J_λ and the contribution of matrix $J_\mu^{0,1,3}$ to connection matrix J_μ, we have

$$
J_\lambda = \left[\begin{array}{cc|cc|cc|cc|cc|cc}
1 & 1 & -1 & 0 & & & 0 & -1 & & & & \\
1 & 1 & -1 & 0 & & & 0 & -1 & & & & \\
\hline
-1 & -1 & 1 & 0 & & & 0 & 1 & & & & \\
0 & 0 & 0 & 0 & & & 0 & 0 & & & & \\
\hline
 & & & & & & & & & & & \\
 & & & & & & & & & & & \\
\hline
0 & 0 & 0 & 0 & & & 0 & 0 & & & & \\
-1 & -1 & 1 & 0 & & & 0 & 1 & & & & \\
\hline
 & & & & & & & & & & & \\
 & & & & & & & & & & & \\
\hline
 & & & & & & & & & & & \\
 & & & & & & & & & & &
\end{array}\right],
$$

$$
J_\mu = \left[\begin{array}{cc|cc|cc|cc|cc|cc}
3 & 1 & -2 & -1 & & & -1 & 0 & & & & \\
1 & 3 & 0 & -1 & & & -1 & -2 & & & & \\
\hline
-2 & 0 & 2 & 0 & & & 0 & 0 & & & & \\
-1 & -1 & 0 & 1 & & & 1 & 0 & & & & \\
\hline
 & & & & & & & & & & & \\
 & & & & & & & & & & & \\
\hline
-1 & -1 & 0 & 1 & & & 1 & 0 & & & & \\
0 & -2 & 0 & 0 & & & 0 & 2 & & & & \\
\hline
 & & & & & & & & & & & \\
 & & & & & & & & & & & \\
\hline
 & & & & & & & & & & & \\
 & & & & & & & & & & &
\end{array}\right].
$$

Let us incorporate the contribution of $\mathrm{T}_1 = \triangle \mathrm{P}_4\mathrm{P}_3\mathrm{P}_1$. For example, the $(0,0)$-th block of matrix $J_\lambda^{4,3,1}$ contributes to the $(4,4)$-th block of matrix J_λ and $(0,2)$-th block of matrix $J_\lambda^{4,3,1}$ contributes to the $(4,1)$-th block of matrix J_λ. Adding the contribution of matrix $J_\lambda^{4,3,1}$ to connection matrix J_λ and the contribution of matrix $J_\mu^{4,3,1}$ to connection matrix J_μ, we have

$$
J_\lambda = \left[\begin{array}{cc|cc|cc|cc|cc|cc}
1 & 1 & -1 & 0 & & & 0 & -1 & & & & \\
1 & 1 & -1 & 0 & & & 0 & -1 & & & & \\
\hline
-1 & -1 & 1 & 0 & & & 0 & 1 & 0 & 0 & & \\
0 & 0 & 0 & 1 & & & 1 & 0 & -1 & -1 & & \\
\hline
 & & & & & & & & & & & \\
 & & & & & & & & & & & \\
\hline
0 & 0 & 0 & 1 & & & 1 & 0 & -1 & -1 & & \\
-1 & -1 & 1 & 0 & & & 0 & 1 & 0 & 0 & & \\
\hline
 & & 0 & -1 & & & -1 & 0 & 1 & 1 & & \\
 & & 0 & -1 & & & -1 & 0 & 1 & 1 & & \\
\hline
 & & & & & & & & & & & \\
 & & & & & & & & & & &
\end{array}\right],
$$

$$
J_\mu = \left[\begin{array}{cc|cc|cc|cc|cc|cc}
3 & 1 & -2 & -1 & & & -1 & 0 & & & & \\
1 & 3 & 0 & -1 & & & -1 & -2 & & & & \\
\hline
-2 & 0 & 3 & 0 & & & 0 & 1 & -1 & -1 & & \\
-1 & -1 & 0 & 3 & & & 1 & 0 & 0 & -2 & & \\
\hline
 & & & & & & & & & & & \\
 & & & & & & & & & & & \\
\hline
-1 & -1 & 0 & 1 & & & 3 & 0 & -2 & 0 & & \\
0 & -2 & 1 & 0 & & & 0 & 3 & -1 & -1 & & \\
\hline
 & & -1 & 0 & & & -2 & -1 & 3 & 1 & & \\
 & & -1 & -2 & & & 0 & -1 & 1 & 3 & & \\
\hline
 & & & & & & & & & & & \\
 & & & & & & & & & & &
\end{array}\right].
$$

Adding block matrices of $J_\lambda^{1,2,4}$ and $J_\lambda^{5,4,2}$ to connection matrix J_λ and block matrices of $J_\mu^{1,2,4}$ and $J_\mu^{5,4,2}$ to connection matrix J_μ, we finally have

$$
J_\lambda = \left[\begin{array}{cc|cc|cc|cc|cc|cc}
1 & 1 & -1 & 0 & & & 0 & -1 & & & & \\
1 & 1 & -1 & 0 & & & 0 & -1 & & & & \\
\hline
-1 & -1 & 2 & 1 & -1 & 0 & 0 & 1 & 0 & -1 & & \\
0 & 0 & 1 & 2 & -1 & 0 & 1 & 0 & -1 & -2 & & \\
\hline
 & & -1 & -1 & 1 & 0 & & & 0 & 1 & 0 & 0 \\
 & & 0 & 0 & 0 & 1 & & & 1 & 0 & -1 & -1 \\
\hline
0 & 0 & 0 & 1 & & & 1 & 0 & -1 & -1 & & \\
-1 & -1 & 1 & 0 & & & 0 & 1 & 0 & 0 & & \\
\hline
 & & 0 & -1 & 0 & 1 & -1 & 0 & 2 & 1 & -1 & -1 \\
 & & -1 & -2 & 1 & 0 & -1 & 0 & 1 & 2 & 0 & 0 \\
\hline
 & & & & 0 & -1 & & & -1 & 0 & 1 & 1 \\
 & & & & 0 & -1 & & & -1 & 0 & 1 & 1
\end{array}\right], \tag{A.30}
$$

$$J_\mu = \left[\begin{array}{rr|rr|rr|rr|rr|rr}
3 & 1 & -2 & -1 & & & -1 & 0 & & & & \\
1 & 3 & 0 & -1 & & & -1 & -2 & & & & \\
\hline
-2 & 0 & 6 & 1 & -2 & -1 & 0 & 1 & -2 & -1 & & \\
-1 & -1 & 1 & 6 & 0 & -1 & 1 & 0 & -1 & -4 & & \\
\hline
& & -2 & 0 & 3 & 0 & & & 0 & 1 & -1 & -1 \\
& & -1 & -1 & 0 & 3 & & & 1 & 0 & 0 & -2 \\
\hline
-1 & -1 & 0 & 1 & & & 3 & 0 & -2 & 0 & & \\
0 & -2 & 1 & 0 & & & 0 & 3 & -1 & -1 & & \\
\hline
& & -2 & -1 & 0 & 1 & -2 & -1 & 6 & 1 & -2 & 0 \\
& & -1 & -4 & 1 & 0 & 0 & -1 & 1 & 6 & -1 & -1 \\
\hline
& & & & -1 & 0 & & & -2 & -1 & 3 & 1 \\
& & & & -1 & -2 & & & 0 & -1 & 1 & 3
\end{array}\right]. \tag{A.31}$$

A.1.4 Formulating 2D Elastic Deformation

This section formulates the elastic deformation of a 2D object. Strain energy of the object is approximated by

$$U = \frac{1}{2}\mathbf{u}_{\mathrm{N}}{}^{\mathrm{T}} K \mathbf{u}_{\mathrm{N}}, \tag{A.32}$$

where K denotes the stiffness matrix. Assume that uniform pressure $\mathbf{p} = [\,p_\xi,\, p_\eta\,]^{\mathrm{T}}$ is exerted over an edge $\mathrm{P}_i\mathrm{P}_j$ of a triangle $\triangle\mathrm{P}_i\mathrm{P}_j\mathrm{P}_k$. Work done by the pressure is then given by

$$W = \int_{\mathrm{P}_i\mathrm{P}_j} \mathbf{p}^{\mathrm{T}}\mathbf{u}\, h\, \mathrm{d}l.$$

Since $\mathbf{u} = N_{i,j,k}\mathbf{u}_i + N_{j,k,i}\mathbf{u}_j$ on edge $\mathrm{P}_i\mathrm{P}_j$, we have

$$\begin{aligned}
W &= \int_{\mathrm{P}_i\mathrm{P}_j} \mathbf{p}^{\mathrm{T}}\{N_{i,j,k}\mathbf{u}_i + N_{j,k,i}\mathbf{u}_j\}\, h\, \mathrm{d}l \\
&= \frac{\mathrm{P}_i\mathrm{P}_j \cdot h}{2}\mathbf{p}^{\mathrm{T}}\mathbf{u}_i + \frac{\mathrm{P}_i\mathrm{P}_j \cdot h}{2}\mathbf{p}^{\mathrm{T}}\mathbf{u}_j \\
&= \begin{bmatrix} (\mathrm{P}_i\mathrm{P}_j \cdot h/2)\,\mathbf{p} \\ (\mathrm{P}_i\mathrm{P}_j \cdot h/2)\,\mathbf{p} \end{bmatrix}^{\mathrm{T}} \begin{bmatrix} \mathbf{u}_i \\ \mathbf{u}_j \end{bmatrix}.
\end{aligned}$$

Namely, work done by external pressure can be described by a linear form of nodal displacement vectors. As a result, internal energy can be described by a quadratic form of nodal displacement vectors. Minimization of the internal energy under geometric constraints yields a system of linear equations on nodal displacement vectors and Lagrange multipliers. Solving the linear equations, we can sketch 2D deformation of an elastic object.

Let us illustrate the formulation of 2D elastic deformation by taking a simple example illustrated in Fig. A.6. The left edge of an elastic object given in Fig. A.5 is fixed on a rigid wall. Uniform pressure $\mathbf{p} = [\, p_\xi,\, p_\eta \,]^{\mathrm{T}}$ is exerted over the right edge of the object. Assume that $h = 1$. Nodal displacement vector is given by

$$\mathbf{u}_{\mathrm{N}} = \begin{bmatrix} \mathbf{u}_0 \\ \mathbf{u}_1 \\ \mathbf{u}_2 \\ \mathbf{u}_3 \\ \mathbf{u}_4 \\ \mathbf{u}_5 \end{bmatrix}.$$

Work done by the pressure exerted over edge $\mathrm{P}_2\mathrm{P}_5$ is described as

$$W = \begin{bmatrix} (\mathrm{P}_2\mathrm{P}_5 \cdot h/2)\,\mathbf{p} \\ (\mathrm{P}_2\mathrm{P}_5 \cdot h/2)\,\mathbf{p} \end{bmatrix}^{\mathrm{T}} \begin{bmatrix} \mathbf{u}_2 \\ \mathbf{u}_5 \end{bmatrix} \triangleq \mathbf{f}_{\mathrm{ext}}^{\mathrm{T}} \mathbf{u}_{\mathrm{N}}$$

where

$$\mathbf{f}_{\mathrm{ext}} = \begin{bmatrix} \mathbf{0} \\ \mathbf{0} \\ (\mathrm{P}_2\mathrm{P}_5 \cdot h/2)\,\mathbf{p} \\ \mathbf{0} \\ \mathbf{0} \\ (\mathrm{P}_2\mathrm{P}_5 \cdot h/2)\,\mathbf{p} \end{bmatrix} = \left[\begin{array}{c} 0 \\ 0 \\ \hline 0 \\ 0 \\ \hline (1/2)\,p_\xi \\ (1/2)\,p_\eta \\ \hline 0 \\ 0 \\ \hline 0 \\ 0 \\ \hline (1/2)\,p_\xi \\ (1/2)\,p_\eta \end{array}\right].$$

Boundary conditions $\mathbf{u}_0 = \mathbf{0}$ and $\mathbf{u}_3 = \mathbf{0}$ are integrated into $A^{\mathrm{T}}\mathbf{u}_{\mathrm{N}} = \mathbf{0}$, where

$$A = \begin{bmatrix} I_{2\times 2} & O \\ O & O \\ O & O \\ O & I_{2\times 2} \\ O & O \\ O & O \end{bmatrix} = \left[\begin{array}{cc|cc} 1 & 0 & 0 & 0 \\ 0 & 1 & 0 & 0 \\ \hline 0 & 0 & 0 & 0 \\ 0 & 0 & 0 & 0 \\ \hline 0 & 0 & 0 & 0 \\ 0 & 0 & 0 & 0 \\ \hline 0 & 0 & 1 & 0 \\ 0 & 0 & 0 & 1 \\ \hline 0 & 0 & 0 & 0 \\ 0 & 0 & 0 & 0 \\ \hline 0 & 0 & 0 & 0 \\ 0 & 0 & 0 & 0 \end{array}\right]$$

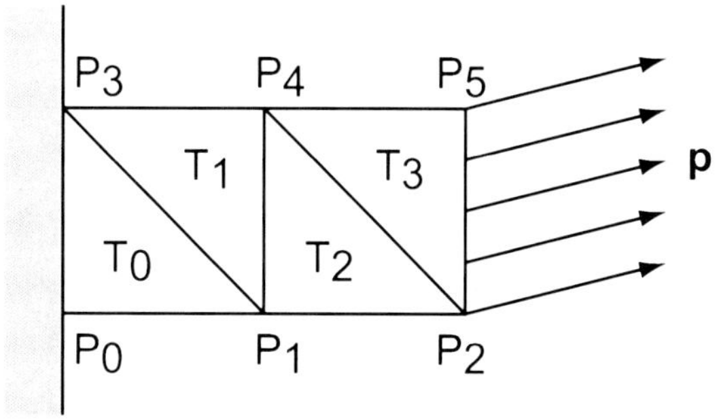

Fig. A.6 Example of 2D elastic object fixed to wall

is a 12×4 matrix. The left half of matrix A denotes the condition $\mathbf{u}_0 = \mathbf{0}$ while its right half corresponds to condition $\mathbf{u}_3 = \mathbf{0}$.

Consequently, the elastic deformation of the object illustrated in Fig. A.6 can be computed by solving the following conditional minimization on nodal displacement vector:

$$\begin{aligned} &\text{minimize} \quad I(\mathbf{u}_\mathrm{N}) = \frac{1}{2}\mathbf{u}_\mathrm{N}{}^\mathrm{T} K \mathbf{u}_\mathrm{N} - \mathbf{f}_\mathrm{ext}^\mathrm{T} \mathbf{u}_\mathrm{N}, \\ &\text{subject to} \quad A^\mathrm{T} \mathbf{u}_\mathrm{N} = \mathbf{0}. \end{aligned}$$

Introducing Lagrange multiplier vector λ_A, which consists of four multipliers, the above conditional minimization turns into an unconditional minimization:

$$\text{minimize} \quad J(\mathbf{u}_\mathrm{N}, \lambda_A) = I(\mathbf{u}_\mathrm{N}) - \lambda_A^\mathrm{T} A^\mathrm{T} \mathbf{u}_\mathrm{N}$$

We can minimize function $J(\mathbf{u}_\mathrm{N}, \lambda_A)$ by solving the following equations:

$$\begin{aligned} \frac{\partial J}{\partial \mathbf{u}_\mathrm{N}} &= K\mathbf{u}_\mathrm{N} - \mathbf{f}_\mathrm{ext} - A\lambda_A = \mathbf{0}, \\ \frac{\partial J}{\partial \lambda_A} &= -A^\mathrm{T} \mathbf{u}_\mathrm{N} = \end{aligned}$$

Namely, the minimization of function $J(\mathbf{u}_\mathrm{N}, \lambda_A)$ turns into a system of linear equations on $\mathbf{u}_\mathrm{N}$ and λ_A:

$$\left[\begin{array}{c|c} K & -A \\ \hline -A^\mathrm{T} & 0_{4\times 4} \end{array}\right] \left[\begin{array}{c} \mathbf{u}_\mathrm{N} \\ \hline \lambda_A \end{array}\right] = \left[\begin{array}{c} \mathbf{f}_\mathrm{ext} \\ \hline \mathbf{0} \end{array}\right].$$

Note that the above linear equation is solvable since the matrix is regular, implying that we can compute $\mathbf{u}_\mathrm{N}$. As a result, we can sketch the elastic deformation of the object.

Note that the first equation describes a set of force equilibrium equations at nodal points. Vector $-K\mathbf{u}_\mathrm{N}$ represents a set of elastic forces applied to the nodal points, $\mathbf{f}_\mathrm{ext}$ and $A\lambda_A$ denote external and constraint forces.

A.1.5 Dynamic 2D Deformation

Dynamic deformation of a 2D object can be formulated by introducing inertial forces. Let us formulate the kinetic energy of the object to derive inertial forces applied to the object. Let ρ be the density of an object at point $\mathrm{P}(\xi, \eta)$. Total kinetic energy of the object is simply given by

$$T = \int_S \frac{1}{2} \rho\, \dot{\mathbf{u}}^{\mathrm{T}} \dot{\mathbf{u}}\, h\, \mathrm{d}S. \tag{A.33}$$

Partitioning region S into a set of triangles, the kinetic energy is described as

$$T = \sum_{\triangle \mathrm{P}_i \mathrm{P}_j \mathrm{P}_k} T_{i,j,k} \tag{A.34}$$

where

$$T_{i,j,k} = \int_{\triangle \mathrm{P}_i \mathrm{P}_j \mathrm{P}_k} \frac{1}{2} \rho\, \dot{\mathbf{u}}^{\mathrm{T}} \dot{\mathbf{u}}\, h\, \mathrm{d}S \tag{A.35}$$

provides partial kinetic energy on region $\triangle \mathrm{P}_i \mathrm{P}_j \mathrm{P}_k$. Velocity of any point within the region can be approximated as follows:

$$\dot{\mathbf{u}} = \dot{\mathbf{u}}_i N_{i,j,k} + \dot{\mathbf{u}}_j N_{j,k,i} + \dot{\mathbf{u}}_k N_{k,i,j}.$$

Substituting the above approximation into Eq. A.35 yields

$$T_{i,j,k} = \frac{1}{2} \left[\, \dot{\mathbf{u}}_i^{\mathrm{T}}\ \dot{\mathbf{u}}_j^{\mathrm{T}}\ \dot{\mathbf{u}}_k^{\mathrm{T}} \,\right] M_{i,j,k} \begin{bmatrix} \dot{\mathbf{u}}_i \\ \dot{\mathbf{u}}_j \\ \dot{\mathbf{u}}_k \end{bmatrix},$$

$$M_{i,j,k} = \begin{bmatrix} M^{i,j,k}{}_{i,i} & M^{i,j,k}{}_{i,j} & M^{i,j,k}{}_{i,k} \\ M^{i,j,k}{}_{j,i} & M^{i,j,k}{}_{j,j} & M^{i,j,k}{}_{j,k} \\ M^{i,j,k}{}_{k,i} & M^{i,j,k}{}_{k,j} & M^{i,j,k}{}_{k,k} \end{bmatrix},$$

where $M^{i,j,k}{}_{i,i}$ through $M^{i,j,k}{}_{k,k}$ are 2×2 matrices given by

$$M^{i,j,k}{}_{i,i} = \left(\int_{\triangle \mathrm{P}_i \mathrm{P}_j \mathrm{P}_k} \rho N_{i,j,k} N_{i,j,k}\, h\, \mathrm{d}S \right) I_{2\times 2},$$

$$M^{i,j,k}{}_{i,j} = \left(\int_{\triangle \mathrm{P}_i \mathrm{P}_j \mathrm{P}_k} \rho N_{i,j,k} N_{j,k,i}\, h\, \mathrm{d}S \right) I_{2\times 2},$$

$$M^{i,j,k}{}_{i,k} = \left(\int_{\triangle \mathrm{P}_i \mathrm{P}_j \mathrm{P}_k} \rho N_{i,j,k} N_{k,i,j}\, h\, \mathrm{d}S \right) I_{2\times 2},$$

and so on.

Assuming that density ρ is constant, computation of partial kinetic energy $T_{i,j,k}$ results in the integration of multiplication of two 2D shape functions. Let $\mathbf{n}$ be a collective vector consisting of shape functions:

$$\mathbf{n} \triangleq \begin{bmatrix} N_{i,j,k} \\ N_{j,k,i} \\ N_{k,i,j} \end{bmatrix}.$$

Let us introduce the following integral:

$$N \triangleq \int_{\triangle \mathrm{P}_i \mathrm{P}_j \mathrm{P}_k} \mathbf{n}\mathbf{n}^{\mathrm{T}}\, h \,\mathrm{d}S.$$

Calculating the above integral (see Appendix C), we have

$$N = \frac{\triangle h}{12} \begin{bmatrix} 2 & 1 & 1 \\ 1 & 2 & 1 \\ 1 & 1 & 2 \end{bmatrix}.$$

Then, matrix $M_{i,j,k}$ is given by multiplying matrix $I_{2\times 2}$ to individual elements of matrix N and multiplying density ρ to the obtained matrix:

$$M_{i,j,k} = \frac{\rho \triangle h}{12} \begin{bmatrix} 2I_{2\times 2} & I_{2\times 2} & I_{2\times 2} \\ I_{2\times 2} & 2I_{2\times 2} & I_{2\times 2} \\ I_{2\times 2} & I_{2\times 2} & 2I_{2\times 2} \end{bmatrix}. \tag{A.36}$$

Note that the sum of all blocks of matrix $M_{i,j,k}$ is equal to $\rho \triangle h I_{2\times 2}$, which denotes the mass of the triangular element. As a result, total kinetic energy can be described by a quadratic form of the nodal velocity vector as follows:

$$T = \frac{1}{2} \dot{\mathbf{u}}_{\mathrm{N}}{}^{\mathrm{T}} M \dot{\mathbf{u}}_{\mathrm{N}}, \tag{A.37}$$

where matrix M is referred to as an *inertia matrix*. A set of inertial forces applied to nodal points is then given by

$$-\frac{\mathrm{d}}{\mathrm{d}t}\frac{\partial T}{\partial \dot{\mathbf{u}}_{\mathrm{N}}} = -\frac{\mathrm{d}}{\mathrm{d}t} M \dot{\mathbf{u}}_{\mathrm{N}} = -M \ddot{\mathbf{u}}_{\mathrm{N}}.$$

Let us illustrate the formulation of dynamic 2D elastic deformation by taking an example illustrated in Fig. A.6. Assume that $h = 1$. Note that the $(0,0)$-th block of matrix $M_{i,j,k}$ contributes to the (i,i)-th block of inertia matrix M, the $(0,1)$-th block of matrix $M_{i,j,k}$ contributes to the (i,j)-th block of the inertia matrix, and so on. Thus, contribution of matrix $M_{0,1,3}$ to inertia matrix is given as follows:

$$\frac{\rho}{24} \begin{bmatrix} 2I_{2\times 2} & I_{2\times 2} & O & I_{2\times 2} & O & O \\ I_{2\times 2} & 2I_{2\times 2} & O & I_{2\times 2} & O & O \\ O & O & O & O & O & O \\ I_{2\times 2} & I_{2\times 2} & O & 2I_{2\times 2} & O & O \\ O & O & O & O & O & O \\ O & O & O & O & O & O \end{bmatrix}.$$

Contribution of matrix $M_{1,4,3}$ to the inertia matrix is then given by

$$\frac{\rho}{24}\begin{bmatrix} O & O & O & O & O & O \\ O & 2I_{2\times2} & O & I_{2\times2} & I_{2\times2} & O \\ O & O & O & O & O & O \\ O & I_{2\times2} & O & 2I_{2\times2} & I_{2\times2} & O \\ O & I_{2\times2} & O & I_{2\times2} & 2I_{2\times2} & O \\ O & O & O & O & O & O \end{bmatrix}.$$

Summing up all contributions, we finally have

$$M = \frac{\rho}{24}\begin{bmatrix} 2I_{2\times2} & I_{2\times2} & O & I_{2\times2} & O & O \\ I_{2\times2} & 6I_{2\times2} & I_{2\times2} & 2I_{2\times2} & 2I_{2\times2} & O \\ O & I_{2\times2} & 4I_{2\times2} & O & 2I_{2\times2} & I_{2\times2} \\ I_{2\times2} & 2I_{2\times2} & O & 4I_{2\times2} & I_{2\times2} & O \\ O & 2I_{2\times2} & 2I_{2\times2} & I_{2\times2} & 6I_{2\times2} & I_{2\times2} \\ O & O & I_{2\times2} & O & I_{2\times2} & 2I_{2\times2} \end{bmatrix}.$$

Lagrangian under geometric constraints $A^{\mathrm{T}}\mathbf{u}_{\mathrm{N}}$ is formulated as

$$\begin{aligned} L &= T - U + W + \lambda_A^{\mathrm{T}} A^{\mathrm{T}}\mathbf{u}_{\mathrm{N}} \\ &= \frac{1}{2}\dot{\mathbf{u}}_{\mathrm{N}}{}^{\mathrm{T}} M \dot{\mathbf{u}}_{\mathrm{N}} - \frac{1}{2}\mathbf{u}_{\mathrm{N}}{}^{\mathrm{T}} K \mathbf{u}_{\mathrm{N}} + \mathbf{f}_{\mathrm{ext}}^{\mathrm{T}}\mathbf{u}_{\mathrm{N}} + \lambda_A^{\mathrm{T}} A^{\mathrm{T}}\mathbf{u}_{\mathrm{N}}, \end{aligned}$$

where λ_A is a set of Lagrange multipliers. Consequently, a set of motion equations of nodal points is formulated as

$$-K\mathbf{u}_{\mathrm{N}} + \mathbf{f}_{\mathrm{ext}} + A\lambda_A - M\ddot{\mathbf{u}}_{\mathrm{N}} = \mathbf{0}. \tag{A.38}$$

Recall that the CSM converts a set of geometric constraints $A^{\mathrm{T}}\mathbf{u}_{\mathrm{N}} = \mathbf{0}$ into a set of differential equations given as

$$A^{\mathrm{T}}\ddot{\mathbf{u}}_{\mathrm{N}} + A^{\mathrm{T}}(2\omega\dot{\mathbf{u}}_{\mathrm{N}} + \omega^2\mathbf{u}_{\mathrm{N}}) = 0. \tag{A.39}$$

Introducing velocity vector $\mathbf{v}_{\mathrm{N}} = \dot{\mathbf{u}}_{\mathrm{N}}$, the equations of motion and the above differential equations can be described as follows:

$$\begin{aligned} \dot{\mathbf{u}}_{\mathrm{N}} &= \mathbf{v}_{\mathrm{N}}, \\ M\dot{\mathbf{v}}_{\mathrm{N}} - A\lambda_A &= -K\mathbf{u}_{\mathrm{N}} + \mathbf{f}_{\mathrm{ext}}, \\ -A^{\mathrm{T}}\dot{\mathbf{v}}_{\mathrm{N}} &= A^{\mathrm{T}}(2\omega\mathbf{v}_{\mathrm{N}} + \omega^2\mathbf{u}_{\mathrm{N}}). \end{aligned} \tag{A.40}$$

Namely,

$$\begin{aligned} \dot{\mathbf{u}}_{\mathrm{N}} &= \mathbf{v}_{\mathrm{N}}, \\ \begin{bmatrix} M & -A \\ -A^{\mathrm{T}} & \end{bmatrix}\begin{bmatrix} \dot{\mathbf{v}}_{\mathrm{N}} \\ \lambda_A \end{bmatrix} &= \begin{bmatrix} -K\mathbf{u}_{\mathrm{N}} + \mathbf{f}_{\mathrm{ext}} \\ A^{\mathrm{T}}(2\omega\mathbf{v}_{\mathrm{N}} + \omega^2\mathbf{u}_{\mathrm{N}}) \end{bmatrix}. \end{aligned} \tag{A.41}$$

Note that the second linear equation is solvable since the matrix in the left hand is regular, implying that we can compute $\dot{\mathbf{v}}_{\mathrm{N}}$. As a result, we can sketch

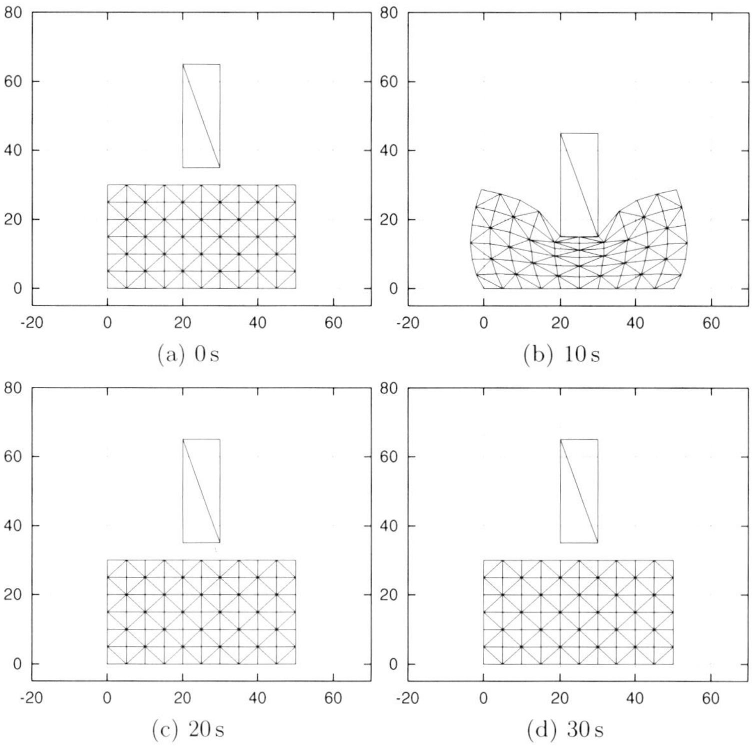

Fig. A.7 Computation of 2D viscoelastic deformation using FEM

$\mathbf{u}_\mathrm{N}$ and $\mathbf{v}_\mathrm{N}$ using any numerical method to solve a set of ordinary differential equations (Appendix B.1).

Note that $-K\mathbf{u}_\mathrm{N}$ describes a set of elastic forces applied to nodal points, where $K = \lambda J_\lambda + \mu J_\mu$. Lamé's constants λ and μ, which characterize the isotropic elasticity, are described as $\lambda = \nu E/(1+\nu)(1-2\nu)$ and $\mu = E/2(1+\nu)$. A set of viscous forces applied to nodal points is then given by $B\dot{\mathbf{u}}_\mathrm{N}$, where $B = \lambda^{\mathrm{vis}} J_\lambda + \mu^{\mathrm{vis}} J_\mu$. Let c be viscous modulus and ν^{vis} be Poisson's ratio for viscosity. Two constants λ^{vis} and μ^{vis}, which characterize the isotropic viscosity, are described as follows:

$$\lambda^{\mathrm{vis}} = \frac{\nu^{\mathrm{vis}} c}{(1+\nu^{\mathrm{vis}})(1-2\nu^{\mathrm{vis}})}, \qquad \mu^{\mathrm{vis}} = \frac{c}{2(1+\nu^{\mathrm{vis}})}.$$

Replacing a set of elastic forces $-K\mathbf{u}_\mathrm{N}$ in Eq. A.41 by a set of viscoelastic forces $-K\mathbf{u}_\mathrm{N} - B\mathbf{v}_\mathrm{N}$, we can compute 2D viscoelastic deformation. Fig. A.7 demonstrates 2D viscoelastic deformation. A rigid bar is pushing down an elastic object fixed on a table. The rigid bar moves downward during the first 10 s and moves upward during the next 10 s. Elastic and viscous moduli are $E = 30$ and $c = 20$. Poisson ratios for the moduli are given by $\nu = 0.35$

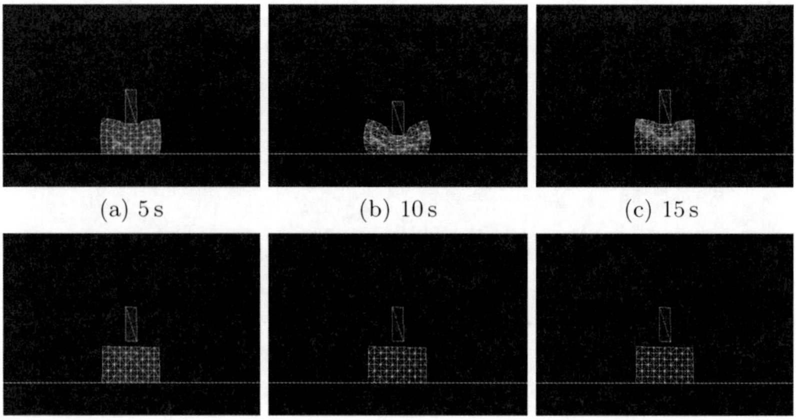

(a) 5 s (b) 10 s (c) 15 s

(d) 20 s (e) 25 s (f) 30 s

Fig. A.8 Computed stress in 2D viscoelastic deformation

Table A.1 Procedure to compute nodal elastic forces

initialization
compute J_λ^p and J_μ^p for each triangle T_p.
compute $K_p = \lambda J_\lambda^p + \mu J_\mu^p$ for each triangle T_p.

elastic force computation
reset $\mathbf{F}_0$ through $\mathbf{F}_{n-1}$.
for p in all triangles
  let P_i, P_j, and P_k be three nodal points of triangle T_p.
  compute $\mathbf{f}_i^p$, $\mathbf{f}_j^p$, and $\mathbf{f}_k^p$ by Eq. A.42.
  decrease $\mathbf{F}_i$ by $\mathbf{f}_i^p$, $\mathbf{F}_j$ by $\mathbf{f}_j^p$, and $\mathbf{F}_k$ by $\mathbf{f}_k^p$.

and $\nu^{\mathrm{vis}} = 0.35$. As shown in the figure, finite element approach can simulate viscoelastic deformation. Fig. A.8 denotes the stress imposed on individual triangles. Note that the stress is relaxed after the contact between the rigid bar and the elastic object is lost.

Computation of Nodal Elastic Forces

In dynamic simulation, we do not have to construct total stiffness matrix K once a set of elastic forces can be computed. Let us compute a set of nodal forces described by $K\mathbf{u}_{\mathrm{N}}$ without constructing total stiffness matrix K or a pair of total connection matrices J_λ and J_μ. Note that a pair of partial connection matrices are determined to each triangle $\mathrm{T}_p = \triangle \mathrm{P}_i\mathrm{P}_j\mathrm{P}_\mathrm{k}$. Instead of specifying three nodal points P_i, P_j, and P_k, indicate triangle T_p in the description of partial connection matrices. Namely, use J_λ^p and J_μ^p instead of $J_\lambda^{i,j,k}$ and $J_\mu^{i,j,k}$. Recall that forces applied to nodal points P_i, P_j, and P_k caused by the elastic deformation of triangle $\mathrm{T}_p = \triangle \mathrm{P}_i\mathrm{P}_j\mathrm{P}_\mathrm{k}$ are specified as follows:

$$\begin{bmatrix} \mathbf{f}_i^p \\ \mathbf{f}_j^p \\ \mathbf{f}_k^p \end{bmatrix} = K_p \begin{bmatrix} \mathbf{u}_i \\ \mathbf{u}_j \\ \mathbf{u}_k \end{bmatrix}, \tag{A.42}$$

where $K_p = \lambda J_\lambda^p + \mu J_\mu^p$. Summing up the contributions of all triangles, we can obtain a set of elastic forces applied to nodal points. In the example illustrated in Fig. A.6,

$$\begin{bmatrix} \mathbf{f}_0^0 \\ \mathbf{f}_1^0 \\ \mathbf{f}_3^0 \end{bmatrix} = K_0 \begin{bmatrix} \mathbf{u}_0 \\ \mathbf{u}_1 \\ \mathbf{u}_3 \end{bmatrix}, \quad \begin{bmatrix} \mathbf{f}_1^1 \\ \mathbf{f}_4^1 \\ \mathbf{f}_3^1 \end{bmatrix} = K_1 K_1 \begin{bmatrix} \mathbf{u}_1 \\ \mathbf{u}_4 \\ \mathbf{u}_3 \end{bmatrix},$$

$$\begin{bmatrix} \mathbf{f}_1^2 \\ \mathbf{f}_2^2 \\ \mathbf{f}_4^2 \end{bmatrix} = K_2 \begin{bmatrix} \mathbf{u}_1 \\ \mathbf{u}_2 \\ \mathbf{u}_4 \end{bmatrix}, K_2 \begin{bmatrix} \mathbf{u}_1 \\ \mathbf{u}_2 \\ \mathbf{u}_4 \end{bmatrix}, \quad \begin{bmatrix} \mathbf{f}_2^3 \\ \mathbf{f}_5^3 \\ \mathbf{f}_4^3 \end{bmatrix} = K_3 \begin{bmatrix} \mathbf{u}_2 \\ \mathbf{u}_5 \\ \mathbf{u}_4 \end{bmatrix}.$$

Then, a set of elastic forces at nodal points is given as follows:

$$\text{nodal elastic forces} = \begin{bmatrix} -\mathbf{f}_0^0 \\ -\mathbf{f}_1^0 - \mathbf{f}_1^1 - \mathbf{f}_1^2 \\ -\mathbf{f}_2^2 - \mathbf{f}_2^3 \\ -\mathbf{f}_3^0 - \mathbf{f}_3^1 \\ -\mathbf{f}_4^1 - \mathbf{f}_4^2 - \mathbf{f}_4^3 \\ -\mathbf{f}_5^3 \end{bmatrix}.$$

As shown above, we do not have to construct total stiffness matrix K or a pair of total connection matrices J_λ and J_μ during the computation of nodal elastic forces. Let $\mathbf{F}_0$ through $\mathbf{F}_{n-1}$ denote elastic forces applied to nodal points P_0 through P_{n-1}. Procedure to compute nodal elastic forces is summarized in Table A.1.

Similarly, we can compute nodal viscous forces without constructing total damping matrix B, implying that we can simulate 2D viscoelastic deformation without constructing K or B.

Lumped Inertia Matrix

In the simulation of dynamic elastic deformation, we have to inverse the inertia the matrix in Eq. A.41. Note that we have assumed that mass is distributed within an object uniformly, which yields non-diagonal elements in inertia matrix M. To avoid the numerical computation of matrix inversion, lumped approximation of inertia can be applied, in which we assume that mass is concentrated to nodal points. In lumped approximation, mass of triangle $\triangle \mathrm{P}_i \mathrm{P}_j \mathrm{P}_k$ is equally distributed to particles P_i, P_j, and P_k. Partial inertia matrix is then described as follows:

$$M_{i,j,k} = \frac{\rho \triangle \mathrm{P}_i \mathrm{P}_j \mathrm{P}_k \cdot h}{3} \begin{bmatrix} I_{2\times 2} & O & O \\ O & I_{2\times 2} & O \\ O & O & I_{2\times 2} \end{bmatrix}.$$

Note that inertia matrix M be diagonal since matrix $M_{i,j,k}$ is diagonal, as shown in the above equation. In the example illustrated in Fig. A.6,

$$M = \frac{\rho h}{6}\begin{bmatrix} I_{2\times 2} & O & O & O & O & O \\ O & 3I_{2\times 2} & O & O & O & O \\ O & O & 2I_{2\times 2} & O & O & O \\ O & O & O & 2I_{2\times 2} & O & O \\ O & O & O & O & 3I_{2\times 2} & O \\ O & O & O & O & O & I_{2\times 2} \end{bmatrix}.$$

Then, the following matrix should be inverted during the simulation of dynamic elastic deformation:

$$\begin{bmatrix} M & -A \\ -A^T & O \end{bmatrix} = \left[\begin{array}{cccccc|cc} m_0 I & & & & & & -I & \\ & m_1 I & & & & & & \\ & & m_2 I & & & & & \\ & & & m_3 I & & & & -I \\ & & & & m_4 I & & & \\ & & & & & m_5 I & & \\ \hline -I & & & & & & & \\ & & & -I & & & & \end{array}\right],$$

where $m_0 = m_5 = \rho h/6$, $m_1 = m_4 = \rho h/2$, and $m_2 = m_3 = \rho h/3$. The inverse of the above matrix can be analytically derived as follows:

$$\begin{bmatrix} M & -A \\ -A^T & O \end{bmatrix}^{-1} = \left[\begin{array}{cccccc|cc} O & & & & & & -I & \\ & m_1' I & & & & & & \\ & & m_2' I & & & & & \\ & & & O & & & & -I \\ & & & & m_4' I & & & \\ & & & & & m_5' I & & \\ \hline -I & & & & & & -m_0 I & \\ & & & -I & & & & -m_3 I \end{array}\right],$$

where $m_1' = 1/m_1$, $m_2' = 1/m_2$, and so on. The above equation suggests that we can simulate the dynamic deformation without constructing total inertia matrix M or numerical computation of an inverse matrix. Actually, we can directly compute $\dot{\mathbf{v}}_0$ through $\dot{\mathbf{v}}_5$ by the following equations:

$$\begin{aligned} \dot{\mathbf{v}}_0 &= -(2\omega \mathbf{v}_0 + \omega^2 \mathbf{u}_0), \\ \dot{\mathbf{v}}_1 &= m_1'(\text{nodal force at } \mathrm{P}_1), \\ \dot{\mathbf{v}}_2 &= m_2'(\text{nodal force at } \mathrm{P}_2), \\ \dot{\mathbf{v}}_3 &= -(2\omega \mathbf{v}_3 + \omega^2 \mathbf{u}_3), \\ \dot{\mathbf{v}}_4 &= m_4'(\text{nodal force at } \mathrm{P}_4), \\ \dot{\mathbf{v}}_5 &= m_5'(\text{nodal force at } \mathrm{P}_5). \end{aligned}$$

As a result, lumped approximation of inertia results in fast computation of the dynamic deformation.

A.2 Summary

Finite element computation follows *divide-and-conquer* approach. Energy over a complex area is divided into energies over simple areas constructing the complex area. Energy over each simple area can be computed analytically through linear approximation of displacement vectors. This chapter employs triangular elements in 2D deformation and tetrahedron elements in 3D deformation. We have obtained analytical description of potential and kinetic energies stored in a triangular element or in a tetrahedron element. Summing up energies over triangular elements or in tetrahedron elements, we have total potential or kinetic energy. Once we have analytical description on potential and kinetic energies, we can apply variational principles in static or in dynamics to compute static/dynamic elastic deformation of an object.

Isotropic 2D/3D elastic deformation can be formulated in a similar way as summarized in Table A.2. Each normal strain component corresponds to an axis while each shear strain component corresponds to a plane, resulting that we have two normal component and one shear component in 2D while three normal component and three shear components in 3D.

Table A.2 Comparison between 2D and 3D deformation

<table>
<tr><th></th><th>2D</th><th>3D</th></tr>
<tr><td>axis
normal strain</td><td>ξ, η
$\varepsilon_{\xi\xi}, \varepsilon_{\eta\eta}$</td><td>$\xi, \eta, \zeta$
$\varepsilon_{\xi\xi}, \varepsilon_{\eta\eta}, \varepsilon_{\zeta\zeta}$</td></tr>
<tr><td>plane
normal strain</td><td>$\xi\eta$
$\varepsilon_{\xi\eta}$</td><td>$\eta\zeta, \zeta\xi, \xi\eta$
$\varepsilon_{\eta\zeta}, \varepsilon_{\zeta\xi}, \varepsilon_{\xi\eta}$</td></tr>
<tr><td>I_λ</td><td>$\left[\begin{array}{cc|c} 1 & 1 & \\ 1 & 1 & \\ \hline & & \end{array}\right]$</td><td>$\left[\begin{array}{ccc|ccc} 1 & 1 & 1 & & & \\ 1 & 1 & 1 & & & \\ 1 & 1 & 1 & & & \\ \hline & & & & & \\ & & & & & \\ & & & & & \end{array}\right]$</td></tr>
<tr><td>I_μ</td><td>$\left[\begin{array}{cc|c} 2 & & \\ & 2 & \\ \hline & & 1 \end{array}\right]$</td><td>$\left[\begin{array}{ccc|ccc} 2 & & & & & \\ & 2 & & & & \\ & & 2 & & & \\ \hline & & & 1 & & \\ & & & & 1 & \\ & & & & & 1 \end{array}\right]$</td></tr>
</table>

B

Numerical Integration of Ordinary Differential Equations

B.1 Runge-Kutta Method

Let us solve an ordinary differential equation

$$\dot{x} = f(x, t)$$

at discrete times $t_n = nT$, where T denotes a constant time interval. The following methods provide an iterative equation that compute $x_{n+1} = x(t_{n+1})$ from $x_n = x(t_n)$. This implies that, starting from the initial value $x_0 = x(0)$, we can obtain the value of x_n using the equation iteratively.

Euler method (one-stage method)

$$x_{n+1} = x_n + Tf(x_n, t_n) \tag{B.1}$$

Heun method (two-stage method)

$$\begin{aligned} x_{n+1} &= x_n + \frac{T}{2}(k_1 + k_2), \\ k_1 &= f(x_n, t_n), \\ k_2 &= f(x_n + Tk_1, t_n + T). \end{aligned} \tag{B.2}$$

Runge-Kutta method (four-stage method)

$$\begin{aligned} x_{n+1} &= x_n + \frac{T}{6}(k_1 + 2k_2 + 2k_3 + k_4), \\ k_1 &= f(x_n, t_n), \\ k_2 &= f(x_n + \frac{1}{2}Tk_1, t_n + \frac{1}{2}T), \\ k_3 &= f(x_n + \frac{1}{2}Tk_2, t_n + \frac{1}{2}T), \\ k_4 &= f(x_n + Tk_3, t_n + T). \end{aligned} \tag{B.3}$$

Runge-Kutta method is illustrated in Fig. B.1. As shown in the figure, k_1 through k_4 are gradients at different pairs of x and t. Increment $x_{n+1} - x_n$ is given by a weighted sum of the four gradients. Noting that $\dot{x}(t_n) = f(x_n, t_n)$, both sides of the formula coincide with each other up to their first order with respect to time interval T in Euler method. Noting that $k_2 = f + T(f_x f + f_t)$ and $\ddot{x}(t_n) = f_x f + f_t$, where $f = f(x_n, t_n)$, $f_x = \partial f/\partial x(x_n, t_n)$, and $f_t = \partial f/\partial t(x_n, t_n)$, both sides of the formula coincide with each other up to their second order in Heun method. It can be shown that both sides of the formula coincide with each other up to their fourth order in Runge-Kutta method. Consequently, Euler method, Heun method, and Runge-Kutta method provide the first order, the second order, and the fourth order solutions, respectively.

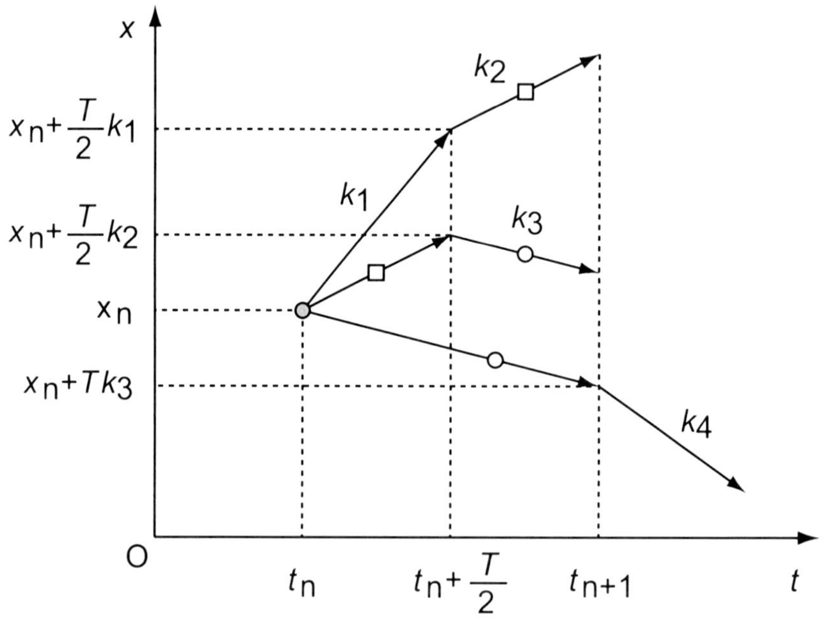

Fig. B.1 Runge-Kutta method

Time interval T should be constant in the above methods. In addition, an inappropriately large time interval may yield an incorrect solution. Thus, we have to select an appropriately small time interval to obtain a correct solution, which results in much computation time. Adaptive selection of time interval has been proposed to improve both solution correctness and time efficiency. The following formula allows us to select an appropriate time interval during the computation.

Runge-Kutta-Fehlberg method (six-stage method)

$$\begin{aligned}
x_{n+1} &= x_n + T(\frac{16}{135}k_1 + \frac{6656}{12825}k_3 + \frac{28561}{56430}k_4 - \frac{9}{50}k_5 + \frac{2}{55}k_6),\\
k_1 &= f(x_n, t_n),\\
k_2 &= f(x_n + \frac{T}{4}k_1, t_n + \frac{1}{4}T),\\
k_3 &= f(x_n + \frac{T}{32}(3k_1 + 9k_2), t_n + \frac{3}{8}T),\\
k_4 &= f(x_n + \frac{T}{2179}(1932k_1 - 7200k_2 + 7296k_3), t_n + \frac{12}{13}T),\\
k_5 &= f(x_n + T(\frac{439}{216}k_1 - 8k_2 + \frac{3680}{513}k_3 - \frac{845}{4104}k_4), t_n + T),\\
k_6 &= f(x_n + T(-\frac{8}{27}k_1 + 2k_2 - \frac{3544}{2565}k_3 + \frac{1859}{4104}k_4 - \frac{11}{40}k_5), t_n + \frac{1}{2}T).
\end{aligned} \tag{B.4}$$

Runge-Kutta-Fehlberg method provides the fifth order solution. The following algorithm, which is referred to as *Runge-Kutta-Fehlberg formula*, updates the time interval adaptively:

Step 1 Compute x_{n+1} using Eq. B.4.
Step 2 Compute x^*_{n+1} given by

$$x^*_{n+1} = x_n + T(\frac{25}{216}k_1 + \frac{1408}{2565}k_3 + \frac{2197}{4104}k_4 - \frac{1}{5}k_5). \tag{B.5}$$

Step 3 Compute $\hat{T}$ given by

$$\hat{T} = \alpha T \left\{ \frac{\epsilon}{\|x^*_{n+1} - x_{n+1}\|} \right\}^{\frac{1}{5}} \tag{B.6}$$

where ϵ is a small positive allowance and α denotes a safety ratio about 0.8 through 0.9.
Step 4 Select time interval T equal to or smaller than $\hat{T}$.

It can be shown that x^*_{n+1} is a fourth order solution of the ordinary differential equation. Using the difference of orders between x_{n+1} and x^*_{n+1}, an appropriate time interval T can be selected adaptively.

Any above method can be applied to a set of differential equations

$$\dot{\mathbf{x}} = \mathbf{f}(\mathbf{x}, t)$$

where $\mathbf{x}$ consists of a set of state variables and $\mathbf{f}$ consists of a set of functions that compute the time derivatives of individual state variables. Replacing state variable x in an above method by state variable vector $\mathbf{x}$, scalar function f by vector function $\mathbf{f}$, and scalar k by vector $\mathbf{k}$ yields a numerical method to integrate a set of ordinary differential equations.

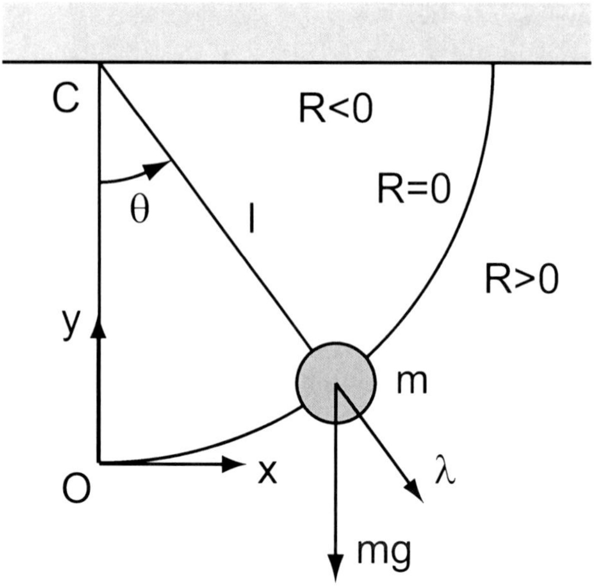

Fig. B.2 Simple pendulum model

B.2 Constraint Stabilization Method (CSM)

Constraint Stabilization Method, which is abbreviated as *CSM*, provides a numerical computation of a system of differential equations under geometric constraints [99]. Let us explain the method by taking the motion equation of a simple pendulum.

Let us investigate the motion of a simple pendulum of length l and mass m suspended from point O, as illustrated in Fig. B.2. Let $\theta(t)$ be the angle of the pendulum at time t. Then, angle θ satisfies the following differential equation:

$$(ml^2)\ddot{\theta} = -mgl\sin\theta$$

Introducing $\omega = \dot{\theta}$, the above differential equation turns into a system of differential equations of the first order:

$$\dot{\theta} = \omega,$$
$$\dot{\omega} = -\frac{g}{l}\sin\theta.$$

Note that no geometric constraints are imposed in this description. Thus, we can simply apply the Euler method or the Runge-Kutta method to solve the above numerically.

Let us formulate the motion of a pendulum in Cartesian coordinates. Let (x, y) be the position of the mass. The kinetic energy T of the pendulum and its potential energy U are described as

$$T = \frac{1}{2}m(\dot{x}^2 + \dot{y}^2), \qquad U = mgy.$$

The following geometric constraint must be satisfied:

$$R(x, y) = \left\{x^2 + (y - l)^2\right\}^{\frac{1}{2}} - l = 0. \tag{B.7}$$

Lagrangian with a geometric constraint is then formulated as follows:

$$\begin{aligned} L &= T - U + \lambda R \\ &= \frac{1}{2}m(\dot{x}^2 + \dot{y}^2) - mgy + \lambda\left[\left\{x^2 + (y - l)^2\right\}^{\frac{1}{2}} - l\right], \end{aligned} \tag{B.8}$$

where λ denotes a Lagrange multiplier, which corresponds to the magnitude of a constraint force since quantity R has the dimension of length. Since R is positive outside the circular trajectory of the pendulum while is negative inside it, force λ represents the outward constraint force as illustrated in the figure. From Eq. B.8, we can derive Lagrange equations of motion:

$$\lambda\, R_x(x, y) - m\ddot{x} = 0, \tag{B.9}$$

$$-mg + \lambda\, R_y(x, y) - m\ddot{y} = 0, \tag{B.10}$$

where $R_x(x, y) = \partial R/\partial x$ and $R_y(x, y) = \partial R/\partial y$. These partial derivatives are given by $R_x(x, y) = xP(x, y)$ and $R_y(x, y) = (y - l)P(x, y)$, where $P(x, y) = (x^2 + (y - l)^2)^{-1/2}$. Note that a geometric constraint Eq. B.7 must be satisfied but simple application of the Euler method or the Runge-Kutta method may break the constraint. We have to incorporate the constraint into the numerical computation of differential equations.

Let us introduce a critical damping of the geometric constraint so that the constraint converges to zero during the computation:

$$\ddot{R} + 2\nu\dot{R} + \nu^2 R = 0, \tag{B.11}$$

where ν is a predetermined angular frequency. Since the above equation provides a critical damping, quantity R converges to zero quickly, say, the geometric constraint is satisfied again during the computation even if the constraint is broken. Substituting Eq. B.7 into the above, we have

$$\begin{aligned} &R_x(x, y)\ddot{x} + R_y(x, y)\ddot{y} + \{\dot{x}^2 + \dot{y}^2\}\, P(x, y) - \{x\dot{x} + (y - l)\dot{y}\}^2 P(x, y)^3 \\ &+2\nu\{x\dot{x} + (y - l)\dot{y}\}\, P(x, y) + \nu^2 R(x, y) = 0. \end{aligned} \tag{B.12}$$

Introducing $v_x = \dot{x}$ and $v_y = \dot{y}$, Eq. B.9, Eq. B.10, and Eq. B.12 are described as follows:

$$\begin{aligned} \dot{x} &= v_x, \\ \dot{y} &= v_y, \\ m\dot{v}_x - R_x(x, y)\,\lambda &= 0, \\ m\dot{v}_y - R_y(x, y)\,\lambda &= -mg, \\ -R_x(x, y)\,\dot{v}_x - R_y(x, y)\,\dot{v}_y &= C(x, y, v_x, v_y) \end{aligned}$$

where

$$C(x, y, v_x, v_y) = (v_x^2 + v_y^2)\, P(x, y) - \{x v_x + (y - l) v_y\}^2 P(x, y)^3 \\ +2\nu\{x v_x + (y - l) v_y\}\, P(x, y) + \nu^2 R(x, y).$$

Consequently, we have the following linear equation on $\dot{x}$, $\dot{y}$, $\dot{v}_x$, $\dot{v}_y$, and λ:

$$\begin{bmatrix} 1\,0 & 0 & 0 & 0 \\ 0\,1 & 0 & 0 & 0 \\ 0\,0 & m & 0 & -R_x(x, y) \\ 0\,0 & 0 & m & -R_y(x, y) \\ 0\,0 & -R_x(x, y) & -R_y(x, y) & 0 \end{bmatrix} \begin{bmatrix} \dot{x} \\ \dot{y} \\ \dot{v}_x \\ \dot{v}_y \\ \lambda \end{bmatrix} = \begin{bmatrix} v_x \\ v_y \\ 0 \\ -mg \\ C \end{bmatrix}$$

Note that the above linear equation is solvable since the matrix is regular, implying that we can compute $\dot{x}$, $\dot{y}$, $\dot{v}_x$, and $\dot{v}_y$. As a result, we can sketch x, y, v_x, and v_y using the Euler method or the Runge-Kutta method. Solving the above linear equation yields

$$\lambda = m\{g R_y(x, y) - C(x, y, v_x, v_y)\},$$
$$\dot{v}_x = \frac{R_x(x, y)}{m}\lambda, \qquad \dot{v}_x = \frac{R_y(x, y)}{m}\lambda - g.$$

From x, y, v_x, and v_y, we can compute λ, then $\dot{v}_x$ and $\dot{v}_y$. Thus, we can apply the Euler method or the Runge-Kutta method to solve the above numerically. The above procedure to incorporate geometric constraints into the numerical solution of dynamical equations is referred to as the constraint stabilization method.

Recall that the gradient vector $[\, R_x,\ R_y \,]^{\mathrm{T}}$ corresponds to the outward normal vector of constraint $R(x, y)$. This vector determines the direction of the constraint force and Lagrange multiplier λ describes its magnitude. Thus, the constraint force is given by $\lambda[\, R_x,\ R_y \,]^{\mathrm{T}}$.

C

Integral over Triangle

Let $P_i(\xi_i, \eta_i)$, $P_j(\xi_j, \eta_j)$, and $P_k(\xi_k, \eta_k)$ be points on $\xi\eta$-plane. Let $O(0,0)$, $A(1,0)$, and $B(0,1)$ be points on $\xi'\eta'$-plane. The following transform converts points O, A, and B into P_i, P_j, and P_k, respectively:

$$\begin{bmatrix} \xi \\ \eta \end{bmatrix} = \begin{bmatrix} \xi_i \\ \eta_i \end{bmatrix} + \begin{bmatrix} \xi_{ji} & \xi_{ki} \\ \eta_{ji} & \eta_{ki} \end{bmatrix} \begin{bmatrix} \xi' \\ \eta' \end{bmatrix}, \tag{C.1}$$

where $\xi_{ji} = \xi_j - \xi_i$, $\xi_{ki} = \xi_k - \xi_i$, $\eta_{ji} = \eta_j - \eta_i$, and $\eta_{ki} = \eta_k - \eta_i$. The above transform can be rewritten as:

$$\begin{bmatrix} 1 \\ \xi \\ \eta \end{bmatrix} = \begin{bmatrix} 1 & 0 & 0 \\ \xi_i & \xi_{ji} & \xi_{ki} \\ \eta_i & \eta_{ji} & \eta_{ki} \end{bmatrix} \begin{bmatrix} 1 \\ \xi' \\ \eta' \end{bmatrix}. \tag{C.2}$$

Since this linear transform converts lines OA, AB, and BA into P_iP_j, P_jP_k, and P_kP_i, respectively, implying that $\triangle$OAB on $\xi'\eta'$-plane is converted into $\triangle P_iP_jP_k$ on $\xi\eta$-plane.

Jacobian corresponding to the above transform is given as:

$$\begin{vmatrix} \partial\xi/\partial\xi' & \partial\xi/\partial\eta' \\ \partial\eta/\partial\xi' & \partial\eta/\partial\eta' \end{vmatrix} = \begin{vmatrix} \xi_{ji} & \xi_{ki} \\ \eta_{ji} & \eta_{ki} \end{vmatrix} = 2\,\triangle P_iP_jP_k.$$

Note that the above determinant is constant; it coincides with the ratio between $\triangle P_iP_jP_k$ and $\triangle$OAB. Let us integrate function $g(\xi,\eta)$ over $\triangle P_iP_jP_k$. Applying the above transform, we have

$$\begin{aligned} \int_{\triangle P_iP_jP_k} g(\xi,\eta)\,\mathrm{d}\xi\,\mathrm{d}\eta &= \int_{\triangle OAB} \hat{g}(\xi',\eta') \det\frac{\partial(\xi,\eta)}{\partial(\xi',\eta')}\,\mathrm{d}\xi'\,\mathrm{d}\eta' \\ &= 2\triangle P_iP_jP_k \int_0^1 \left\{ \int_0^{1-\xi'} \hat{g}(\xi',\eta')\,\mathrm{d}\eta' \right\} \mathrm{d}\xi', \end{aligned}$$

where

$$\hat{g}(\xi', \eta') = g(\xi_i + \xi_{ji}\xi' + \xi_{ki}\eta', \eta_i + \eta_{ji}\xi' + \eta_{ki}\eta').$$

Let $N_{i,j,k}(\xi, \eta)$ be a basic function within $\triangle \mathrm{P}_i\mathrm{P}_j\mathrm{P}_k$ that takes 1 at P_i while 0 at P_j and P_k. Assume that $N_{i,j,k}(\xi, \eta)$ be given in a linear form as:

$$N_{i,j,k}(\xi, \eta) = a_i + b_i\xi + c_i\eta = \begin{bmatrix} a_i \; b_i \; c_i \end{bmatrix} \begin{bmatrix} 1 \\ \xi \\ \eta \end{bmatrix}.$$

Coefficients a_i, b_i, and c_i must satisfy

$$\begin{bmatrix} a_i \; b_i \; c_i \end{bmatrix} \begin{bmatrix} 1 & 1 & 1 \\ \xi_i & \xi_j & \xi_k \\ \eta_i & \eta_j & \eta_k \end{bmatrix} = \begin{bmatrix} 1 \; 0 \; 0 \end{bmatrix}.$$

Solving the above equation, we can determine coefficients a_i, b_i, and c_i. Assume that $N_{i,j,k}(\xi, \eta)$ corresponds to $\hat{N}_{i,j,k}(\xi', \eta')$ through transform Eq. (C.2). Then,

$$\begin{aligned} \hat{N}_{i,j,k}(\xi', \eta') &= N_{i,j,k}(\xi_i + \xi_{ji}\xi' + \xi_{ki}\eta', \eta_i + \eta_{ji}\xi' + \eta_{ki}\eta') \\ &= \begin{bmatrix} a_i \; b_i \; c_i \end{bmatrix} \begin{bmatrix} 1 & 0 & 0 \\ \xi_i & \xi_{ji} & \xi_{ki} \\ \eta_i & \eta_{ji} & \eta_{ki} \end{bmatrix} \begin{bmatrix} 1 \\ \xi' \\ \eta' \end{bmatrix}. \end{aligned}$$

Noting that

$$\begin{bmatrix} 1 & 0 & 0 \\ \xi_i & \xi_{ji} & \xi_{ki} \\ \eta_i & \eta_{ji} & \eta_{ki} \end{bmatrix} = \begin{bmatrix} 1 & 1 & 1 \\ \xi_i & \xi_j & \xi_k \\ \eta_i & \eta_j & \eta_k \end{bmatrix} \begin{bmatrix} 1 & -1 & -1 \\ 0 & 1 & 0 \\ 0 & 0 & 1 \end{bmatrix},$$

we have

$$\begin{aligned} \hat{N}_{i,j,k}(\xi', \eta') &= \begin{bmatrix} a_i \; b_i \; c_i \end{bmatrix} \begin{bmatrix} 1 & 1 & 1 \\ \xi_i & \xi_j & \xi_k \\ \eta_i & \eta_j & \eta_k \end{bmatrix} \begin{bmatrix} 1 & -1 & -1 \\ 0 & 1 & 0 \\ 0 & 0 & 1 \end{bmatrix} \begin{bmatrix} 1 \\ \xi' \\ \eta' \end{bmatrix} \\ &= \begin{bmatrix} 1 \; 0 \; 0 \end{bmatrix} \begin{bmatrix} 1 & -1 & -1 \\ 0 & 1 & 0 \\ 0 & 0 & 1 \end{bmatrix} \begin{bmatrix} 1 \\ \xi' \\ \eta' \end{bmatrix} \\ &= 1 - \xi' - \eta'. \end{aligned}$$

Let $N_{j,k,i}(\xi, \eta) = a_j + b_j\xi + c_j\eta$ be basic function that takes 1 at P_j and 0 at P_k and P_i while $N_{k,i,j}(\xi, \eta) = a_k + b_k\xi + c_k\eta$ be basic function that takes 1 at P_k and 0 at P_i and P_j. Coefficients must satisfy

$$\begin{bmatrix} a_j \; b_j \; c_j \end{bmatrix} \begin{bmatrix} 1 & 1 & 1 \\ \xi_i & \xi_j & \xi_k \\ \eta_i & \eta_j & \eta_k \end{bmatrix} = \begin{bmatrix} 0 \; 1 \; 0 \end{bmatrix},$$

$$\begin{bmatrix} a_k \; b_k \; c_k \end{bmatrix} \begin{bmatrix} 1 & 1 & 1 \\ \xi_i & \xi_j & \xi_k \\ \eta_i & \eta_j & \eta_k \end{bmatrix} = \begin{bmatrix} 0 \; 0 \; 1 \end{bmatrix}.$$

Table C.1 Integral of multiplication of two 2D shape functions

	$N_{i,j,k}$	$N_{j,k,i}$	$N_{k,i,j}$
$N_{i,j,k}$	$\triangle/6$	$\triangle/12$	$\triangle/12$
$N_{j,k,i}$	$\triangle/12$	$\triangle/6$	$\triangle/12$
$N_{k,i,j}$	$\triangle/12$	$\triangle/12$	$\triangle/6$

$(\triangle = \triangle \mathrm{P}_i \mathrm{P}_j \mathrm{P}_k)$

Assuming that $N_{j,k,i}(\xi, \eta)$ and $N_{k,i,j}(\xi, \eta)$ correspond to $\hat{N}_{j,k,i}(\xi', \eta')$ and $\hat{N}_{k,i,j}(\xi', \eta')$, we find that

$$\hat{N}_{j,k,i}(\xi', \eta') = \xi', \qquad \hat{N}_{k,i,j}(\xi', \eta') = \eta'.$$

We can then calculate integral of multiplication of two basic functions over $\triangle \mathrm{P}_i \mathrm{P}_j \mathrm{P}_k$. For example,

$$\begin{aligned} \int_{\triangle \mathrm{P}_i \mathrm{P}_j \mathrm{P}_k} N_{j,k,i} N_{k,i,j} \, \mathrm{d}\xi \, \mathrm{d}\eta &= 2 \triangle \mathrm{P}_i \mathrm{P}_j \mathrm{P}_k \int_0^1 \left\{ \int_0^{1-\xi'} \xi' \eta' \, \mathrm{d}\eta' \right\} \mathrm{d}\xi' \\ &= \frac{1}{12} \triangle \mathrm{P}_i \mathrm{P}_j \mathrm{P}_k. \end{aligned}$$

Table C.1 summarizes the integral of multiplication of two basic functions in $\triangle \mathrm{P}_i \mathrm{P}_j \mathrm{P}_k$.

References

1. Morley, J.W.: Neural Apects of Tactile Sensation. Elsevier Science B. V. (1998)
2. Okamura, A.M., Smaby, N., Cutkosky, M.T.: An Overview of Dexterous Manipulation. In: Proc. of IEEE International Conference on Robotics and Automations Symposium (2000)
3. Dupont, P.E.: Friction Modeling in Dynamic Robot Simulation. In: Proc. of IEEE International Conference on Robotics and Automations Symposium, pp. 1370–1376 (1990)
4. Dahl, P.: A Solid Friction Model. Tech. Rep. TOR-0158(3107-18) (1968)
5. Canudas, C., Olsson, H., Astrom, K.J.: Lischinsky A New Model for Control of Systems with Friction. IEEE Transactions on Automatic Control 40, 419–425 (1995)
6. Haessig, D.A., Friedland, B.: On the Modeling and Simulation of Friction. In: Proc. American Control Conference, vol. 1, pp. 1256–1261 (1990)
7. Xydas, N., Bhagavat, M., Kao, I.: Study of soft-finger contact mechanics using finite analysis and experiments. In: Proc. of IEEE International Conference on Robotics and Automations, vol. 3, pp. 2179–2184 (2000)
8. Maeno, T., Hiromitsu, S., Kawai, T.: Control of Grasping Force by Detecting Stick/Slip Distribution at the Curved Surface of an Elastic Finger. In: Proc. of IEEE International Conference on Robotics and Automations, pp. 3895–3900 (2000)
9. Nguyen, P.T.A., Arimoto, S.: Computer simulation of controlled motion of dual fingers with soft tips grasping and manipulating an object. Advanced Robotics 16, 123–145 (2002)
10. Inoue, T., Hirai, S.: Elastic Model of Deformable Fingertip for Soft-fingered Manipulation. IEEE Transaction on Robotics 22, 1273–1279 (2006)
11. Ghafoor, A., Dai, J.S., Duffy, J.: Stifness Modeling of the Soft-Finger Contact in Robotic Grasping. Journal of Mechanical Design 126, 646–656 (2004)
12. Kao, I., Cutkosky, M.R.: Quasistatic Manipulation with Compliance and Sliding. The Int. Jour. of Robotic Research 11, 20–40 (1992)
13. Fearing, R.S.: Implementing a force strategy for object reorientation. In: Proc. of IEEE International Conference on Robotics and Automations, vol. 3, pp. 96–102 (1986)
14. Peshkin, M.A., Sanderson, A.C.: Manipulation of Sliding Object. In: Proc. of IEEE International Conference on Robotics and Automations, vol. 1, pp. 233–239 (1986)

15. Brock, D.L.: On Operators with a Separable Spectrum. In: Proc. of IEEE International Conference on Robotics and Automations, vol. 1, pp. 249–251 (1988)
16. Kappagantu, R.V., Feeny, B.F.: Analysis and Modeling of an Experimental Frictionally Excited Beam. In: Guran, A., Pfeiffer, F., Popp, K. (eds.) Dynamics with Friction: Modeling, Analysis and Experimental Part II. World Scientific Publishing Company, Singapore (2001)
17. Tremblay, M.R., Cutkosky, M.R.: Estimating Friction Using Incipient Slip Sensing During a Manipulation Task. In: Proc. of IEEE International Conference on Robotics and Automations, vol. 1, pp. 429–434 (1993)
18. Son, J.S., Monteverde, E.A., Howe, R.D.: A Tactile Sensor for Localizing Transient Events in Manipulation. In: Proc. of IEEE International Conference on Robotics and Automations, vol. 1, pp. 471–476 (1994)
19. Mukaibo, Y., Shirado, H., Konyo, M., Maeno, T.: Development of a Texture Sensor Emulating the Tissue Structure and Perceptual Machanism of Human Fingers. In: Proc. of IEEE International Conference on Robotics and Automations, pp. 2565–2570 (2005)
20. Engel, J., Chen, J., Wang, X., Fan, Z., Liu, C., Jones, D.: Technology Development of Integrated Multi-modal And Flexible Tactile Skin For Robotics Applications. In: Proc. IEEE/RSJ Int. Conf. on Robotics System, vol. 3, pp. 2359–2364 (2003)
21. Beccai, L., Roccella, S., Ascari, L., Vandastri, P., Sieber, A., Corrozza, M.C., Dario, P.: Development and Experimental Analysis of a Soft Compliant Tactile Microsensor for Anthropormorphic Artificial Hand. IEEE Transaction on Mechatronics 12, 158–168 (2008)
22. Boissieu, F.: Tactile Texture Recognition with a 3-Axial Force. MEMS integrated Artificial Finger. Robotics: Science and System V (2009)
23. Kim, K.: A silicon-based flexible tactile sensor for ubiquitous robot companion applications. J. Phys. Conf. Ser. 34, 399–403 (2006)
24. Vasarhelyi: Characterisation of an integrable single-crystalline 3-D Tactile Sensor. IEEE Sensor Journal 6, 928–934 (2006)
25. Johnson, J.L.: Contact Mechanics. Cambridge University Press (1984)
26. Levesque, V., Hayward, V.: Experimental evidence of lateral skin strain during tactile exploration. In: Proc. 2003 Eurohaptics, pp. 261–275 (2003)
27. Tada, M.: Individual Difference in Contact. Journal of the Robotics Society of Japan 30, 17–19 (2012)
28. Timoshenko, S., Goodier, J.N.: Theory of Elasticity. McGraw-Hill (1970)
29. Yoshikawa, T., Kurisu, M.: Identification of the Center of Friction from Pushing an Object by a Mobile Robot. In: Proc. of IEEE/RSJ Int. Workshop on Intelligent Robot and Systems, pp. 449–454 (1991)
30. Lynch, M.K.: Estimating the Friction Parameters of Pushed Objects. In: Proc. of IEEE/RSJ Int. Conf. on Intelligent Robot and Systems, pp. 186–193 (1993)
31. Baraff, D.: Analytical Methods for Dynamic Simulation of Non Penetrating Rigid Bodies. SIGGRAPH Compute Graph 23, 223–232 (1989)
32. Hippmann: An algorithm for compliant contact between complexly shaped bodies. Multibody System Dynamics 12, 345–362 (1989)
33. Barbagli, F., Frisoli, A., Salisbury, K., Bergamasco, M.: Simulating human fingers: a Soft Finger Proxy Model and Algorithm. In: 12th International Symposium on Haptic Interfaces for Virtual Environment and Teleoperator Systems, pp. 9–17 (2004)

34. Mindlin, R.D.: Compliance of elastic bodies in contact. Trans. ASME, J. Appl. Mech. 16, 259–268 (1949)
35. Minamizawa, K.: Displaying Grasping and Weight Sensation Based on Cutaneous Stimuli. Journal of the Robotics Society of Japan 30, 39–41 (2012)
36. Nakatani, M., Kawasoe, T.: Haptic Skill Logger (Haplog): The Wearable Sensor for Evaluating Haptic Behaviors. Journal of the Robotics Society of Japan 30, 47–49 (2012)
37. Konyo, M., Okamoto, S.: Pseudo Haptic Representations Using Vibrotactile Stimuli. Journal of the Robotics Society of Japan 30, 23–25 (2012)
38. Nahvi, A., Hollerback, J.M., Freier, R., Nelson, D.D.: Display of Friction in Virtual Environments Based on Human Finger Pad Charactersistics. In: Proc. of ASME Dynamic Systems and Control Division, pp. 179–184 (1998)
39. Astrom, K.J., Canuda-de-wit, C.: Revisiting the Lugre Friction Model. IEEE Control Systems Magazine, 101–114 (2008)
40. Srinivasan, M.A., Dandekar, K.: An Investigation of the Mechanics od Tactile Sense Using Two-Dimensional Models of the Promate Fingertip. ASME Journal of Biomechanical Engineering 118, 48–55 (1996)
41. Shimawaki, S., Sakai, N.: Quasi-static Deformation Analysis of a Human Finger Using a Thee-dimensional Finite Element Model Constructed from CT Images. Journal of Environment and Engineering 2, 56–63 (2007)
42. Tada, M., Shibata, T., Imai, M., Osagawara, T.: Development of the Simulataneous Measurement System of Finger Tip Deformation and Grip/Load Force for Studies of Human Grasping Skill. Transactions of the Japan Institute of Electronics, Information and Communication Enginners D-II 84, 1033–1044 (2001)
43. Wang, J.K., Wang, L., Ho, V.A., Morikawa, S., Hirai, S.: A 3D Non-Homogeneous FE Model of Human Fingertip Based on MRI Measurements. IEEE Transaction on Instrumentation and Measurement 61, 3147–3157 (2012)
44. Jalon, J., Bayo, E.: Kinetic and Dynamic Simulation of Multibody Systems. Springer (1993)
45. Lee, M.H.: Tactile Sensing: New Directions, New Challenges. International Journal of Robotics Research 19, 636–643 (2000)
46. Bradsky, G., Kaehler, A.: Learning OpenCV. O Reilly Press (2008)
47. Thrun, S., Burgard, W., Fox, D.: Probabilistic Robotics. MIT Press (2005)
48. Gray, R.: Entropy and Information Theory. Springer (1990)
49. Hertenstein, M.J., Weiss, S.J.: The Handbook of Touch. Springer (2011)
50. Tada, M., Kanade, T.: An Imaging System of Incipient Slip for Modeling How Human Perceives Slip of a Fingertip. In: Proc. of 26th Annual International Conference of the IEEE EMBS, pp. 2045–2048 (2004)
51. Schmitz, A., Maiolino, P., Maggiali, M., Natale, L., Catana, G., Metta, G.: Methods and Technologies for the Implementation of Large-Scale Robot Tactile Sensors. IEEE Transactions on Robotics 27, 389–400 (2011)
52. Dao, D.V., Toriyama, T., Wells, J.C., Sugiyama, S.: Micro Force-moment Sensor with Six-Degree of Freedom. In: IEEE Int. Sym. on Micromechatronics and Human Science, pp. 93–98 (2001)
53. Dao, D.V.: Ltudy on Silicon Piezoresisive Six-Degree of Freedom Micro Force-Moment Sensors and Application to Fluid Mechanics. PhD Thesis, Ritsumeikan University, Kyoto (2003)
54. Hosoda, K.: Anthropomorphic robotics soft fingertip with randomly distributed receptors. Robot. Autom. Syst. 54, 104–109 (2006)

55. Fujii, I., Inoue, T., Dao, D.V., Sugiyama, S., Hirai, S.: Tactile Perception Using Micro Force/Moment Sensor Embedded in Soft Fingertip. In: Proc. IEEE Sensors, pp. 558–562 (2006)
56. Mittendorfer, P., Cheng, G.: Humanoid Multimodal Tactile-Sensing Modules. IEEE Transactions on Robotics 27, 400–409 (2011)
57. Pezzementi, Z., Jantho, E., Estrade, L., Hager, G.D.: Characterization and Simulation of Tactile Sensor. In: Haptics Symposium, pp. 199–205 (2010)
58. Kidono, K., Seki, H., Kuroda, N., Kamiya, Y., Hikizu, M.: Static Characteristic of Tension-Sensitive Conductive Yarn. In: CD-R Proceedings of The 22th Annual Conference of The Robotics Society of Japan, pp. 1–4 (2006)
59. Paradiso, R., Loriga, G., Taccini, N.: Wearable Health Care System for Vital Signs Monitoring. In: Proc. of 4th International IEEE EMBS Special Topic Conference on Information Technology Applications in Biomedicine, pp. 283–286 (2003)
60. Hasegawa, Y., Shikida, M., Ogura, D., Sata, K.: Glove Type of Wearable Tactile Sensor Produced by Artificial Hollow Fiber. In: Transducer and EuroSensor, pp. 283–286 (2003)
61. Tawil, D.S., Rye, D., Velonaki, M.: Improved Image Reconstruction for an EIT-Based Sensitive Skin with Multiple Internal Electrodes. IEEE Transactions on Robotics 27, 425–435 (2011)
62. Scilingo, E.P., Gemigani, A., Paradiso, R., Taccini, N., Ghelarducci, B., De Rossi, D.: Performance Evaluation of Sensing Fabrics for Monitoring Physiological and Biomechanical Variables. IEEE Transactions on Information Technology in Biomedicine 9, 345–352 (2005)
63. Lukowicz, P., Troster, G.: Textile Pressure Sensor for Muscle Activity and Motion Detection. In: Proc. IEEE Int. Symp. Wearable Comput., pp. 69–75 (2006)
64. Lorussi, F., Galatolo, S., Bartalesi, R., De Rossi, D.: Modeling and Characterization of Extensible Wearable Textile-Based Electrogoniometers. IEEE Sensors Journal 13, 217–228 (2012)
65. Veltink, P.H., De Rossi, D.: Wearable Technology for Biomechanics: e-Textile or Micromechanical Sensors. IEEE Eng. Med. Biol. Magazine, 37–43 (2010)
66. Buscher, G., Koiva, R., Schurmann, C., Haschke, R., Ritter, H.J.: Flexible and Stretchable Fabric-based Tactile Sensor. In: 2012 IEEE Intl. Conf. on Intelligent Robots and System Workshop on Advances in Tactile Sensing and Touch-based Human-Robot Interaction (2012)
67. Lumelsky, V., Shur, M., Wagner, S.: Sensitive Skins: Electronics on Flexible Substrates. IEEE Sensors Journal 1, 37–43 (2001)
68. Son, J.S., Cutkosky, M., Howe, R.: Comparison of Contact Sensor Localization Abilities During Manipulation. In: Proc. IEEE/RSJ Int. Conf. on Robots and System, pp. 96–103 (1995)
69. Teshigawara, S., Tadakuma, K., Ming, A., Ishikawa, M., Shimojo, M.: High Sensitivity Initial Slip Sensor for Dexterous Grasp. In: Proc. of IEEE International Conference on Robotics and Automations, pp. 4867–4872 (2010)
70. Nitta Corp., `http://www.nitta.co.jp`
71. Jamali, N., Sammut, C.: Majority Voting: Material Classification by Tactile Sensing Using Surface Texture. IEEE Transactions on Robotics 27, 508–522 (2011)
72. Oddo, C.M., Controzzi, M., Beccai, L., Cipriani, C., Carrozza, M.C.: Vibrotactile Recognition and Categorization of Surfaces by a Humanoid Robot. IEEE Transactions on Robotics 27, 523–533 (2011)

73. Giguere, P., Dudek, G.: A Simple Tactile Probe for Surface Identification by Mobile Robots. IEEE Transactions on Robotics 27, 534–544 (2011)
74. Weeks, M.: Digital Signal Processing Using Matlab and Wavelets. Infinity Science Press LLC (2007)
75. Okamura, A.M., Turner, M.L., Cutkosky, M.R.: Haptic Exploration of Objects with Rolling and Sliding. In: Proc. of IEEE International Conference on Robotics and Automations Symposium, pp. 2485–2490 (1997)
76. Lepora, N.F., Evans, M., Fox, C.W., Diamond, E., Gurney, K., Prescott, T.J.: Naive Bayes Texture Classification Applied to Whisker Data from a Moving Robot. In: The 2010 International Joint Conference on Neural Networks, pp. 1–8 (2010)
77. Anderson, C.W., Stolz, E.A., Shamsunder, S.: Multivariate Autoregressive Models for Classification of Spontaneous Electroencephalographic Signals during Mental Tasks. IEEE Transactions on Biomedical Engineering 45, 277–286 (1998)
78. Tsoi, A.C., So, D.S., Sergejew, A.: Classification of Electroencephalographic Using Artificial Neural Network. Advances in Neural Information Processing System 6, 1151–1158 (1998)
79. Guo, L., Rivero, D., Seoane, J.A., Pazos, A.: Classification of EEG signals using relative wavelet energy and artificial neural networks. In: Proc. of GEC 2009 Proceedings of the First ACM/SIGEVO Summit on Genetic and Evolutionary Computation, pp. 177–183 (2009)
80. Mitchell, T.: Artificial Neural Network, ch 4. McGraw-Hill (1997)
81. Mathworks Inc., `http://www.mathworks.com`
82. Russell, R.: Compliant-Skin Tactile Sensor. In: Proc. of IEEE/RSJ Int. Conf. on Intelligent Robot and Systems, pp. 1645–1648 (1987)
83. Fearing, R.S.: Compliant-Skin Tactile Sensor. In: Proc. of IEEE/RSJ Int. Conf. on Intelligent Robot and Systems, pp. 1637–1643 (1987)
84. Schneider, A.: Object Identification with Tactile Sensors using Bag-of-Features. In: Proc. IEEE/RSJ Int. Conf. on Robots and System, pp. 243–248 (2009)
85. Hutchings, B.L., Peterson, R.J.: Real-Time Control Using Tactile Feedback and Fuzzy Logic. In: Proc. of IEEE World Congress on Computational Intelligence, pp. 1326–1331 (1994)
86. Maekawa, H., Tanie, K., Komoriya, K.: Tactile Sensor Based Manipulation of an Unknown Object by a Multifingered Hand with Rotation Contact. In: Proc. of IEEE International Conference on Robotics and Automation, pp. 743–750 (1995)
87. Dang, H., Weisz, J., Allen, P.K.: Blind Grasping: Stable Robotic Grasping Using Tactile Feedback and Hand Kinematics. In: Proc. IEEE Int. Conf. on Robotics and Automation, pp. 5917–5922 (2011)
88. Bekiroglu, Y., Laaksonen, J., Jorgensen, J.A., Kyrki, V., Kragic, D.: Assessing Grasp Stability Based on Learning and Haptic Data. IEEE Transactions on Robotics 27, 616–629 (2011)
89. Maekawa, H., Tanie, K., Komoriya, K.: Tactile Sensor Based Manipulation of an Unknown Object by a Multifingered Hand with Rotation Contact. In: Proc. IEEE International Conference on Robotics and Automation, pp. 743–750 (1995)
90. Chitta, S., Sturm, J., Piccoli, M., Burgard, W.: Tactile Sensing for Mobile Manipulation. IEEE Transactions on Robotics 27, 558–568 (2011)

91. Kawasaki, H., Komatsu, T., Uchiyama, K.: Dexterous Anthropomorphic Robot Hand With Distributed Tactile Sensor: Gifu Hand II. IEEE/ASME Transaction on Mechatronics 7, 296–303 (2002)
92. Holweg, E.G.M.: Slip Detection by Tactile Sensors: Algorithms and Experimental Results. In: Proc. IEEE International Conference on Robotics and Automation, pp. 3234–3239 (1996)
93. Alcazar, J.A., Barajas, G.: Estimating Object Grasp Sliding via Pressure Array Sensing. In: Proc. IEEE International Conference on Robotics and Automation, pp. 1740–1746 (2012)
94. Hillis: A High-Resolution Imaging Touch Sensor. International Journal of Robotics Research 1, 33–44 (1982)
95. Overton, K.J., Williams, T.: Tactile Sensation for Robots. In: Proc. Seventh Intl Joint Conf. on Artificial Intelligence, pp. 791–795 (1981)
96. Westling, G., Johansson, R.S.: Responses in Glabrous Skin Mechanoreceptors during Precision Grip in Humans. Experimental Brain Research 66, 128–140 (1986)
97. Lowe, D.: Distinctive Image Features from Scale-Invariant Keypoints. Computer Vision 60, 91–140 (2004)
98. Bay, H., Tuytelaars, T., Gool, L.V.: SURF: Speeded Up Robist Features. In: 9th European Conference on Computer Vision, pp. 404–417 (2006)
99. Baumgarte: Stabilization of Constraints and Integrals of Motion in Dynamical Systems. Computer Methods in Applied Mechanics and Engineering 1, 1–16 (1972)

Index

Printed by Publishers' Graphics LLC